Techniques of
VACUUM ULTRAVIOLET
SPECTROSCOPY

PROFESSOR THEODORE LYMAN
(1874–1954)

Reproduced by kind permission of Dr. W. F.
Meggers to whom Lyman presented this
autographed photo in 1926.

VIKTOR SCHUMANN
(1841–1913)

Reproduced by kind permission of the
Lyman Laboratory of Physics, Harvard
University.

Techniques of

VACUUM ULTRAVIOLET

SPECTROSCOPY

JAMES A. R. SAMSON, B.Sc., M.S., Ph.D., D.Sc.
University of Nebraska–Lincoln

Pied Publications, Lincoln, Nebraska

Library of Congress Catalog Card Number: 67–19780
International Standard Book Number: 0–918626–15–3

Second Printing: 1980

Published by:
PIED PUBLICATIONS
1927 South 26 Street
Lincoln, Nebraska 68502

Manufactured in the United States of America

To Mary, Ross, and Scott

Preface

Although the energy range covered by the vacuum ultraviolet region of the spectrum, 6 to 6000 eV, greatly exceeds that of the combined infrared, visible, and ultraviolet regions, no text in English since that written by Lyman in 1928 has been devoted to vacuum ultraviolet radiation. In the preface to his second edition in 1928 Professor Lyman wrote, "In the last fourteen years, the extreme ultraviolet has changed from a little-known region to a well-recognized and important part of the spectrum." Indeed, a great deal of knowledge of atomic and molecular structure was to be obtained by studies in this spectral region. A very active period in vacuum ultraviolet research, however, began about 14 years ago. This activity was largely stimulated by upper-atmosphere and space research. During this period major advances have been made in the techniques and instrumentation necessary to solve many problems. With the surge of interest in this region there has been a need for a book to describe vacuum ultraviolet radiation.

It is the purpose of this book to discuss in some detail the basic elements used in vacuum ultraviolet spectroscopy and to bring together in a central location the results of all the major advances in technique and instrumentation. The book is aimed at the student and researcher entering into this spectral region. It is also intended to be a useful reference book for those already working with vacuum ultraviolet radiation. To this end, references to original papers have been made as extensive as possible.

My thanks are owed to Professor G. L. Weissler, who introduced me to this field and provided a stimulating and enthusiastic atmosphere for vacuum ultraviolet research. It is with pleasure that I acknowledge his helpful suggestions, along with those of Professor W. R. S. Garton and Dr. R. B. Cairns, who have read the entire manuscript, and to Mr. R. W. Hunter for his comments on Chapter 9. Also, I should like to express my appreciation to the GCA Corporation for contributing the necessary secretarial help and to Miss Marion Kachadorian for her careful typing of the manuscript.

I am grateful to the editors of *The Physical Review, Journal of the Optical Society of America, The Review of Scientific Instruments, Journal de Physique et le Radium, Journal of Applied Physics, Japanese Journal of Applied Physics, Applied Optics, Journal of Chemical Physics, Physical Review Letters, Science of Light, Journal of Physics and Chemistry of Solids, Bulletin of the Academy of Sciences of the USSR, Physical Series, Planetary and Space Science, Journal of Scientific Instruments, Annales de Geophysique, Journal of Geophysical Research, Instruments and Experimental Techniques, Space Sciences Reviews, Comptes Rendus, Advances in Geophysics, Optik, Optica Acta, Report on Progress in Physics,* the McGraw-Hill Book Company, and Academic Press, Inc., for permission to reproduce various tables and figures, and also to the individual authors for their permission, credit for which is given in each case by reference to the authors under the figures and tables.

I would like to thank the following companies for supplying either photographs of their equipment or permission to reproduce research results: McPherson Instrument Corporation, Jarrell-Ash Corporation, Bausch and Lomb Company, Piolet Chemicals, Inc., Nuclear Enterprises Ltd., and Harshaw Chemicals, Inc.

JAMES A. R. SAMSON

Waltham, Mass.
April 1967

Contents

Techniques of

VACUUM ULTRAVIOLET
SPECTROSCOPY

ERRATA

page

15. The numbers on the ordinate axis of Fig. 2.5 should be raised one division.

26. In Fig. 2.12 the horizontal arrow represents the arrow marked on the back of a commercial grating. The arrow head must point towards the entrance slit.

154. In Fig. 5.53 the spectral line 1185.7 NI should be 1085.7 NII.

174. The reference in Fig. 5.68 should read [148].

215. The decrease in the efficiency of sodium salicylate between 200 and 300 $\overset{\circ}{A}$, shown in Fig. 7.6, is an artifact of the measurements. The efficiency remains relatively constant between 100 and 400 $\overset{\circ}{A}$.

239. In Fig. 7.37, third line of the caption, $\gamma(\theta^{\circ})$ should read $\gamma(0^{\circ})$.

268. In all the equations the parameters L_1 and L_2 should appear in the exponential terms, that is, $\exp(-\mu L_1)$, etc.

I

Introduction

Owing to the high absorptance of air, early spectroscopic studies in the ultraviolet region of the spectrum were limited to wavelengths longer than about 2000 Å.

In 1893 Viktor Schumann [1] built the first vacuum spectrograph and made the first investigation of vacuum ultraviolet radiation. He employed a fluorite prism as the dispersing element, but because the dispersion of fluorite was unknown, he was unable to determine the wavelengths of the dispersed radiation. Nevertheless, he was able to show that air and, in particular, oxygen was responsible for the absorption below 2000 Å. Theodore Lyman [2], using a vacuum spectrograph equipped with a concave diffraction grating, was the first to measure wavelengths in this region. He found that Schumann's spectrum had a short wavelength limit of about 1250 Å, the limit due to the transmission characteristics of fluorite. The region from 2000 to 1250 Å is now known as the *Schumann region*.

The use of the concave diffraction grating as the dispersing element in a vacuum spectrograph together with the absence of windows enables the entire vacuum ultraviolet region to be investigated down to about 5 Å. Thus the vacuum uv region overlaps the soft x-ray region. The difference between the two regions is simply that ultraviolet or optical radiation corresponds to energy changes of the *outer* electrons of an atom or ion, while x-radiation corresponds to energy changes of the *inner* electrons. The optical radiation in the overlapping region is generally produced from highly ionized atoms, for example, the 13.44 Å resonance line of Ne IX corresponding to the transition $1\,^1S - 2\,^1P$ [3]. The shortest optical emission line presently observed lies at 10.8 Å, while the shortest x-ray line observed with a grating spectrograph lies at 4.7 Å [4]. In the future, lines of shorter wavelengths may be observed. The characteristic $L_{2,3}$ emission band of aluminum at about 180 Å and the beryllium K line at 115 Å are examples of soft x-rays.

Various names have been given to the wavelength region below 2000 Å. However, because of the opacity of air between about 2 and 2000 Å and the consequent need to evacuate spectrographs, this region will be called,

1

following Boyce [5], the *vacuum ultraviolet*. The major divisions of optical radiation are as follows: infrared, wavelengths longer than 7000 Å; visible, 7000 to 4000 Å; ultraviolet, 4000 to 2000 Å; and the vacuum ultraviolet, 2000 to 2 Å. The vacuum ultraviolet region covers a very large energy range, 6 eV to 6 keV; however, it can be conveniently subdivided owing to the nature of the optical instruments used. For example, below 1040 Å no window materials transmit radiation, and below about 300 Å the low reflectance of gratings necessitates the use of a grazing incidence spectrograph. Too, the use of different light sources tends to establish separate regions for the experimentaler. The H_2 glow discharge,

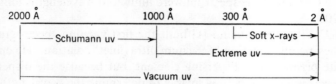

Fig. 1.1 The wavelength range and nomenclature used in the vacuum ultraviolet.

which provides useful radiation from above 2000 Å down to about 900 Å, is frequently used in the Schumann region. At shorter wavelengths, different light sources must be used. Generally these are of the high voltage condensed spark discharge type. This is the region of the *extreme* ultraviolet. Since this region includes the soft x-ray region, it has been suggested by Tousey [6] that it be contracted to XUV. Figure 1.1 shows a chart of the subdivisions and the nomenclature introduced above. The limits should not be taken as precise boundaries.

As in many branches of science, pioneer workers in the vacuum uv had to invent or improve existing instrumentation before they could proceed with their investigations. Schumann [7] improved the photographic plate so that spectra could be recorded which previously had been absorbed by the gelatin in which the emulsion was suspended. Rowland discovered the focusing properties of the concave diffraction grating and techniques were evolved for ruling gratings [8,9].

Compared with the more accessible region of the spectrum, work in the vacuum uv progressed slowly. However, in the last decade interest in this area has rapidly increased largely because of increased activity in atmospheric and space research and in plasma physics. Research has also been stimulated by the availability of excellent commercial vacuum spectrographs and monochromators. Such increased activity in vacuum uv spectroscopy has embraced many branches of science, reviews of which have been given by Weissler [10] on the photoionization and photoelectric phenomena and by Watanabe [11] on atmospheric absorption processes.

As a result of this increased activity and interest, the First International Conference on Vacuum Ultraviolet Radiation Physics was held in 1962 at the University of Southern California. The proceedings of this conference [12] provide an excellent picture of the diverse fields utilizing vacuum uv radiation in their researches. An excellent historical review of the field of vacuum uv spectroscopy has been written by Tousey [13]. Also of historical interest are the works of Bomke [14] and Lyman [15]. In addition to [5] and [10–15], other recent reviews on vacuum uv research and techniques are listed in the general bibliography.

It is the design of this book to present the techniques currently employed in vacuum uv spectroscopy in a form readily available to the researcher and the student.

GENERAL BIBLIOGRAPHY

Boyd, R. L. F., Techniques for the Measurement of Extra-Terrestrial Soft X-Rays, in *Space Science Reviews*, ed. C. de Jager (Reidel, Dordrecht, Holland, 1965), Vol. 4, pp. 35–90.

Edlén, B., Wavelength Measurements in the Vacuum Ultraviolet, in *Reports on Progress in Physics* (The Institute of Physics and the Physical Society, London, 1963), Vol. 26, p. 181.

Garton, W. R. S., Spectroscopy in the Vacuum Ultraviolet, in *Advances in Atomic and Molecular Physics*, eds. D. R. Bates and I. Estermann (Academic, New York, 1966), Vol. 2, pp. 93–176.

Henke, B. L., Measurement in the 10 to 100 Angstrom X-Ray Region, *Advances in X-Ray Analysis* **4**, 244–279 (1961).

Henke, B. L., Production, Detection, and Application of Ultrasoft X-Rays, in *X-Ray Optics and X-Ray Microanalysis*, eds. H. H. Pattee, V. E. Cosslett, and A. Engstrom (Academic, New York, 1963), pp. 157–172.

Inn, E. C. Y., Vacuum Ultraviolet Spectroscopy, *Spectrochimica Acta* **7**, 65–87 (1955).

Mayer, U., Optical Focusing with Soft X-Rays and in the Extreme Ultraviolet, *Space Science Reviews* **3**, 781–815 (1964).

McNesby, J. R., and H. Okabe, Vacuum Ultraviolet Photochemistry, in *Advances in Photochemistry*, eds. W. A. Noyes, G. S. Hammond, and J. N. Pitts (Interscience, New York, 1964), Vol. 3, pp. 157–240.

Price, W. C., Vacuum Ultraviolet Spectroscopy, in *Advances in Spectroscopy*, ed. H. W. Thompson (Academic Press, New York, 1961), Vol. 1, p. 56.

Samson, J. A. R., The Measurement of the Photoionization Cross Sections of the Atomic Gases, in *Advances in Atomic and Molecular Physics*, eds. D. R. Bates and I. Estermann (Academic, New York, 1966), Vol. 2, pp. 177–261.

Samson, J. A. R., and G. L. Weissler, Photon Interactions with Particles, in *Methods of Experimental Physics*, eds. B. Bederson and W. L. Fite (Academic, New York, 1967), Vol. 8.

Samson, J. A. R., Vacuum Ultraviolet Research, *Applied Optics* (March 1967).

Sandström, A. E., Experimental Methods of X-Ray Spectroscopy: Ordinary Wavelengths, in *Handbuch der Physik*, ed. S. Flügge (Springer-Verlag, Berlin, 1957), Vol. 30, pp. 78–245.

Tomboulian, D. H., Experimental Methods of Soft X-Ray Spectroscopy and the Valence Band Spectra of the Light Elements, in *Handbuch der Physik*, ed. S. Flügge (Springer-Verlag, Berlin, 1957), Vol. 30, pp. 246–304.

Vodar, B., *Extreme Ultraviolet Spectroscopy, Proceedings Xth Colloquium Spectroscopium Internationale*, eds. E. R. Lippincott and M. Margoshes (Spartan Books, Washington, D. C., 1963,) pp. 217–246.

Weissler, G. L., Photoionization in Gases and Photoelectric Emission from Solids, in *Handbuch der Physik*, ed. S. Flügge (Springer-Verlag, Berlin, 1956), Vol. 21, pp. 304–382.

Welford, W. T., Aberration Theory of Gratings and Grating Mountings, in *Progress in Optics*, ed. E. Wolf (North-Holland, Amsterdam, 1965), Vol. 4, pp. 243–280.

Wilkinson, P. G., Molecular Spectra in the Vacuum Ultraviolet, *J. Mol. Spectroscopy* **6**, 1–57 (1961).

REFERENCES

[1] V. Schumann, *Akad. Weiss. Wien.* **102**, 2A, 625 (1893).

[2] T. Lyman, *Astrophys. J.* **5**, 349 (1906).

[3] B. C. Fawcett, A. H. Gabriel, B. B. Jones, and N. J. Peacock, *Proc. Phys. Soc.* **84**, 257 (1964).

[4] A. H. Gabriel, J. R. Swain, and W. A. Waller, *J. Sci. Instr.* **42**, 94 (1965).

[5] J. C. Boyce, "Spectroscopy in the Vacuum Ultraviolet," *Revs. Mod. Phys.* **13**, 1–57 (1941).

[6] R. Tousey, *J. Opt. Soc. Am.* **52**, 1186 (1962).

[7] V. Schumann, *Ann. d. Phy.* **5**, 349 (1901).

[8] H. A. Rowland, *Phil. Mag.* **13**, 469 (1882).

[9] H. A. Rowland, *Phil. Mag.* **16**, 197 and 210 (1883).

[10] G. L. Weissler, "Handbuch der Physik" (Springer-Verlag, Berlin, 1956), Vol. XXI.

[11] K. Watanabe, *Advances in Geophysics* **5**, 153–223 (1958).

[12] "Proc. First Intern. Conference Vac. UV Radn. Phys.," ed. G. L. Weissler, *J. Quant. Spectrosc. Radiat. Transfer* **2**, 313 (1962).

[13] R. Tousey, *Appl. Optics* **1**, 679 (1962).

[14] H. Bomke, "Vakuumspektroskopie" Johann Ambrosius Barth, Leipzig, 1937).

[15] T. Lyman, "The Spectroscopy of the Extreme Ultraviolet" (Longmans, Green, New York, 1928) 2nd ed.

2

The Concave Diffraction Grating

In 1882 Prof. H. A. Rowland conceived the idea of combining the principle of the plane diffraction grating with the focusing properties of a concave mirror. He found that such concave gratings had the following excellent but simple focusing properties. If a concave grating is placed tangentially to a circle of a diameter equal to the radius of curvature of the grating such

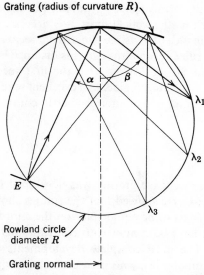

Fig. 2.1 The Rowland circle. Radiation from the point E is dispersed and focused by the grating at λ_1, λ_2, etc. α and β are the angles of incidence and diffraction, respectively.

that the grating center lies on the circumference, the spectrum of an illuminated point lying on the circle will be focused on this circle (see Fig. 2.1). This circle is known as the Rowland circle and forms the basis of nearly all vacuum spectrographs today.

Rowland further contributed to the use of the concave grating by perfecting the ruling engine. Before this, the ruling of plane gratings was

5

difficult and often unsuccessful. In his own words Rowland writes [1]:

"One of the problems to be solved in making a machine is to make a perfect screw; and this mechanics of all countries have sought to do for over a hundred years, and have failed. On thinking over the matter, I devised a plan whose details I shall soon publish, by which I hoped to make a practically perfect screw; and so important did the problem seem, that I immediately set Mr. Schneider, the instrument-maker of the University, at work at once."

The "plan" succeeded, and the near perfect screw was made. As a result, Rowland was able to rule excellent gratings with as many as 43,000 lines per inch. His ruling engine is still the prototype of all such machines in operation today. A complete description of Rowland's ruling engine was given by J. S. Ames [2] in 1902. A more recent discussion on ruling engines and the production of diffraction gratings has been given by Harrison [3], who suggested that ruling engines be interferometrically controlled in order to maintain uniformly spaced grooves. For the resolving power of a grating to approach its theoretical limit, it is necessary for a given groove to be ruled to within a $\frac{1}{10}$ of a visible fringe from its proper position relative to the first groove ruled, even although this distance is 20 cm or more. In 1955, Harrison and Stroke [4] reported the first successful use of inter-ferometric control of a ruling engine. An excellent review of high resolution spectroscopy and ruling under interferometric control has been given by Stroke [5].

2.1   THEORY

The theory of the concave grating was developed to a large degree by Rowland himself [6]. He showed that the rulings should be so spaced on the spherical surface as to be equidistant on the chord of the circular arc.

Since 1883 there have been numerous contributions to the theory of the concave grating. The most complete developments have been given by Beutler [7] and Namioka [8], who used geometrical optics, and by Mack, Stehn, and Edlén [9] who used physical optics. A rather complete biblio-graphy is given in the papers of Namioka.

With the principles of geometrical optics it is possible to determine the focusing properties of the concave grating as well as the degree of astig-matism present. Setting up a Cartesian coordinate system with the origin 0 located at the center of the grating rulings, let the x axis be the grating normal with the z axis parallel to the rulings. Let $A(x, y, z)$, $B(x', x', z')$, and $P(u, w, l)$ be points on the entrance slit, image, and grating, respectively (see Fig. 2.2). Then the condition for two rays reflected from adjacent grooves to reinforce at B is that their path difference must equal an integral number of wavelengths. That is, the path difference $= m\lambda$, where $m = 1, 2, 3, \ldots$, etc., is called the spectral order. The path difference for two rays reflected from grooves a distance w apart with a constant groove separation d is equal to $m\lambda w/d$. Thus for reinforcement of

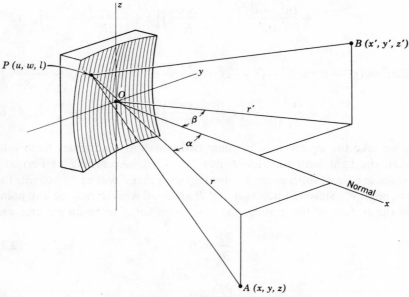

Fig. 2.2 Image formation by the concave grating.

all rays at B any arbitrary light path APB, where P varies over the grating surface, may be represented by the path function F given by

$$F = AP + BP + \frac{wm\lambda}{d}, \qquad (2.1)$$

where

$$(AP)^2 = (x - u)^2 + (y - w)^2 + (z - l)^2, \qquad (2.2)$$

and

$$(BP)^2 = (x' - u)^2 + (y' - w)^2 + (z' - l)^2. \qquad (2.3)$$

Introducing cylindrical coordinates, we see from Fig. 2.2 that $x = r \cos \alpha$, $y = r \sin \alpha$, $x' = r \cos \beta$, and $y' = r' \sin \beta$, where α and β are the angles of incidence and diffraction, respectively. The signs of α and β are opposite if A and B lie on different sides of the xz plane. Furthermore, since all points such as P lie on a sphere of radius R, we have

$$u = R \pm [R^2 - (w^2 + l^2)]^{1/2}. \qquad (2.4)$$

Only the minus sign is significant in this application. Now converting to cylindrical coordinates and substituting eq. (2.4) for u in eqs. (2.2) and (2.3), we find

$$(AP)^2 = (r - w \sin \alpha)^2 + (z - l)^2 - l^2 \frac{r \cos \alpha}{R} + w^2 \left(\cos^2 \alpha - \frac{r \cos \alpha}{R} \right)$$

$$+ \frac{(w^2 + l^2)^2}{4R^2} \left(1 - \frac{r \cos \alpha}{R} \right) \left(1 + \frac{w^2 + l^2}{2R^2} + \cdots \right), \quad (2.5)$$

and

$$(BP)^2 = (r' - w \sin \beta)^2 + (z' - l)^2 - l^2 \frac{r' \cos \beta}{R} + w^2 \left(\cos^2 \beta - \frac{r' \cos \beta}{R} \right)$$

$$+ \frac{(w^2 + l^2)^2}{4R^2} \left(1 - \frac{r' \cos \beta}{R} \right) \left(1 + \frac{w^2 + l^2}{2R^2} + \cdots \right). \quad (2.6)$$

If we take the square root of these two equations and insert them into (2.1), the light path function F can be found. According to Fermat's principle of least time, point B is located such that F will be an extreme for any point P. Since the points A and B are fixed while P may be any point on the surface of the grating, the conditions for F to be an extreme are

$$\frac{\partial F}{\partial l} = 0 \qquad (2.7)$$

and

$$\frac{\partial F}{\partial w} = 0. \qquad (2.8)$$

If (2.7) and (2.8) could be satisfied simultaneously by any pair of l and w for the fixed point B, then B would be the point of perfect focus. However, a perfect image cannot be obtained from a concave grating. As in the case of a spherical concave mirror, the concave grating will image a point source first into a vertical line (horizontal focus), then into a horizontal line (vertical focus). It can be shown from (2.7) and (2.8) that for B to be the best horizontal focal point, the following condition must hold:

$$\frac{\cos^2 \alpha}{r} - \frac{\cos \alpha}{R} + \frac{\cos^2 \beta}{r'} - \frac{\cos \beta}{R} = 0. \qquad (2.9)$$

Two solutions of this equation are

$$r = R \cos \alpha, \qquad r' = R \cos \beta, \qquad (2.10)$$

and

$$r = \infty, \qquad r' = \frac{R \cos^2 \beta}{\cos \alpha + \cos \beta}. \qquad (2.11)$$

Equation 2.10 is the equation of a circle expressed in polar coordinates and is known as the *Rowland circle*. It expresses the fact that diffracted light of all wavelengths will be focused horizontally on the circumference of a circle of diameter R equal to the radius of curvature of the grating, provided that the entrance slit and grating are located on the circle and the grating normal lies along a diameter (see Fig. 2.1). This is the normal condition for observing a spectrum. With vertical entrance slits and vertical rulings it is important that the spectrum be focused in the horizontal plane.

The vertical or secondary foci are more important in the reduction or elimination of astigmatism (see Sect. 2.5). The locus of the vertical foci is given by the equation

$$\frac{1}{r} - \frac{\cos \alpha}{R} + \frac{1}{r'} - \frac{\cos \beta}{R} = 0. \tag{2.12}$$

The solutions of this equation are

$$r = \frac{R}{\cos \alpha}, \qquad r' = \frac{R}{\cos \beta}, \tag{2.13}$$

and

$$r = \infty, \qquad r' = \frac{R}{\cos \alpha + \cos \beta}. \tag{2.14}$$

Equation 2.13 is the equation for a straight line tangent to the Rowland circle at the normal to the grating. Thus any point on the tangent will be focused vertically and brought to a horizontal astigmatic line on the same tangent.

Equations 2.11 and 2.14 represent the case where the incident light is parallel. If the spectrum is viewed near the normal, then $\cos \beta \sim 1$, both these equations become identical, and $r' = R/(1 + \cos \alpha)$. That is, the vertical and horizontal foci coincide, and the image will be stigmatic. This is the condition for the Wadsworth mounting. When the grating is illuminated close to normal, $r' \sim R/2$.

It can also be shown from (2.7) and (2.8) that for the central ray AOB,

$$\left(1 + \frac{z^2}{r^2}\right)^{-\frac{1}{2}} (\sin \alpha + \sin \beta_0) = \frac{m\lambda}{d}, \tag{2.15}$$

and

$$\frac{z}{r} = -\frac{z'_0}{r'_0}, \tag{2.16}$$

where (r'_0, β_0, z'_0) are the coordinates of the image point for the central ray. Equations 2.15 and 2.16 represent, respectively, the grating equation and the geometrical relation between object point and image point. By letting z tend to zero in (2.15) we obtain the same grating equation as for a plane grating. In practice, however, $z^2/r^2 \ll 1$ and can be neglected. Thus the grating equation for a concave grating can be taken as

$$\pm m\lambda = d(\sin \alpha + \sin \beta), \qquad (2.17)$$

where the negative sign applies when the spectrum lies between the central image ($\alpha = \beta$) and the tangent to the grating (sometimes referred to as the "outside order"). When the spectrum lies between the incident beam and the central image, the positive sign must be used, and the spectrum is referred to as the "inside order".

2.2 DISPERSION

The dispersion of a grating expresses how the various wavelengths are distributed along the Rowland circle. The *angular dispersion* is defined as $d\beta/d\lambda$. This quantity can be easily determined by differentiating (2.17). For a fixed angle of incidence we have

ANGULAR DISPERSION $$\frac{d\beta}{d\lambda} = \frac{m}{d \cos \beta}. \qquad (2.18)$$

More frequently we are interested in the actual number of angstroms per mm dispersed along the Rowland circle. This quantity, which is actually the reciprocal of the *linear dispersion* $dl/d\lambda$, is called the *plate factor*. If we rewrite the plate factor in terms of $\Delta\beta$, we have

$$\frac{\Delta\lambda}{\Delta l} = \frac{\Delta\lambda}{\Delta\beta} \cdot \frac{\Delta\beta}{\Delta l}. \qquad (2.19)$$

From Fig. 2.3 we see that $R\Delta\beta = \Delta l$, where R is the radius of the concave grating. Therefore, (2.19) becomes

$$\frac{\Delta\lambda}{\Delta l} = \frac{1}{R(\Delta\beta/\Delta\lambda)}. \qquad (2.20)$$

We can replace Δ in the limit by the differential d. Thus from (2.18) and (2.20), we obtain

PLATE FACTOR $$\frac{d\lambda}{dl} = \frac{d \cos \beta}{mR}$$

$$= \frac{\cos \beta}{mR(1/d)} \times 10^4 \quad \text{Å/mm}, \qquad (2.21)$$

when R is measured in meters and $(1/d)$ is the number of lines per mm.

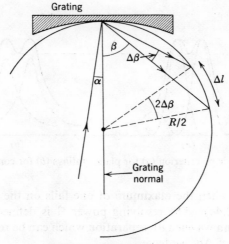

Fig. 2.3 Angular dispersion.

For a 1 m normal incidence spectrograph with a 600 line per mm grating, the plate factor is approximately 16 Å per mm in the first order. The smaller the numerical value of the plate factor, the larger the dispersion. From (2.18), we can see that the dispersion increases for the higher orders and as the number of ruled lines per unit length increase. Near the normal to the grating, $\beta \sim 0$, and the dispersion is nearly constant. However, as β increases, so does the dispersion.

2.3 RESOLVING POWER

The resolving power and dispersion are closely related quantities. While the dispersion determines the separation of two wavelengths, the resolving power determines whether this separation can be distinguished. Each monochromatic beam itself forms a diffraction pattern, the principal maxima of which are represented by the order number m. Between such maxima, secondary maxima exist whose intensities decrease as the number of ruled lines N exposed to the incident radiation increase. In practice, these secondary maxima are very much weaker than the principal maxima. The angular half–width of a principal maximum $\Delta\beta$ is the angular distance between the principal maximum and its first minimum and in a plane grating is given by

$$\Delta\beta = \frac{\lambda}{Nd \cos \beta}.$$ (2.22)

This width then provides a theoretical limit to the resolving power of a grating. If we use Rayleigh's criterion, two lines of equal intensity will

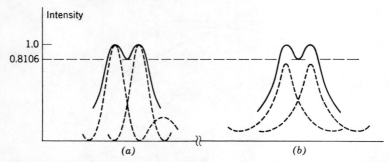

Fig. 2.4 Rayleigh criterion (*a*) for plane gratings (*b*) for concave gratings.

just be resolved when the maximum of one falls on the minimum of the other (see Fig. 2.4*a*). The resolving power $\Re$ is defined as $\lambda/\Delta\lambda$, where $\Delta\lambda$ is the minimum wavelength separation which can be resolved. Expressing $\Delta\lambda$ in terms of $\Delta\beta$, we have,

$$\Delta\lambda = \Delta\beta \cdot \frac{d\lambda}{d\beta},$$

which from (2.18) and (2.22) becomes

$$\Delta\lambda = \frac{\lambda}{mN}.$$

Thus

RESOLVING POWER $$\Re = \frac{\lambda}{\Delta\lambda} = mN. \qquad (2.23)$$

Whether this theoretical limit is achieved will depend on the actual quality of the grating. That the resolving power increases with the order number *m* is clear from the fact that the angular half width of the principal maxima is practically independent of *m*, whereas the angular dispersion essentially increases linearly with *m*. The cos β term in these two relations varies slowly with *m*.

Equation 2.23 is true strictly for the plane diffraction grating. However, it is also true for concave gratings when the width of the grating is less than a certain optimum value. It has been shown by Namioka [8] and by Mack, Stehn, and Edlén [9] that for a concave grating, the angular half-width of a principal maximum deviates from that given by (2.22), and that the diffracted minima do not reach zero. Thus they introduced a modified Rayleigh's criterion. The basic change from Rayleigh's definition is that they do not require the maximum of one line to fall on the first minimum of the other line. However, the ratio of the minimum intensity of the composite structure, as shown in Fig. 2.4*b*, to that of either maximum

Table 2.1 Resolving Power of the Concave
Grating as a Function of Grating Width

Width of Grating W	Resolving Power $\mathcal{R}$
$W \leq W_{\text{opt}}/1.18$	$mN = W(m/d)$
$W = W_{\text{opt}}$	$0.92W_{\text{opt}}(m/d)$
$W \gg W_{\text{opt}}$	$0.75W_{\text{opt}}(m/d)$

remains the same as in the Rayleigh definition, namely, $8/\pi^2$. Expressed formally, the modified criterion states that two lines of equal intensity will be just resolved when the wavelength difference between them is such that the minimum total intensity between the lines is $8/\pi^2$ ($= 0.8106$) times as great as the total intensity at the central maximum of either of the lines. Using this new criterion, they have shown that the resolving power of a concave grating is equal to mN when the width W of the grating illuminated is less than or equal to $W_{\text{opt}}/1.18$. The optimum width W_{opt} is discussed below. As the width of the grating increases beyond $W_{\text{opt}}/1.18$, the resolving power still increases (but not so rapidly as mN) until it reaches a maximum at a width $W = W_{\text{opt}}$. This optimum resolving power, $\mathcal{R}_{\text{opt}}$, is then equal to $0.92W_{\text{opt}}(m/d)$. For $W > W_{\text{opt}}$, the resolving power tends to oscillate about a mean value such that for $W \gg W_{\text{opt}}$, $\mathcal{R} = 0.75W_{\text{opt}}(m/d)$. The above results are summarized in Table 2.1, and typical values of W_{opt} are listed in Table 2.2. The above discussion applies to the case when the center of the illuminated slit lies on the Rowland circle, known as in-plane mounting. For off-plane mountings, such as the Eagle mounting, $\mathcal{R}_{\text{opt}} = 0.95W_{\text{opt}}(m/d)$ (see Chapter 3).

Table 2.2 Optimum Width W_{opt} (cm) for a 1-Meter Grating
(for gratings with radius R cm multiply W_{opt} by $[R/100]^{3/4}$)

	$m/d = 6000$ cm^{-1}						$m/d = 12000$ cm^{-1}				
λ(Å) α^0	10	100	500	1000	2000	λ(Å) α^0	10	100	500	1000	2000
0	57.5	32.4	21.6	18.2	15.3	0	40.7	22.9	15.3	12.8	10.7
7	3.4	6.1	9.6	12.0	15.2	7	3.4	6.2	10.1	12.7	12.8
35	1.5	2.7	4.0	4.9	5.9	35	1.5	2.7	4.1	5.0	6.2
45	1.3	2.3	3.5	4.2	5.1	45	1.3	2.3	3.5	4.3	5.4
80	0.8	1.4	2.2	2.7	3.3	80	0.8	1.4	2.2	2.7	3.3
88	0.5	1.0	1.6	1.9	2.3	88	0.5	1.0	1.6	1.9	2.3
90	0	0	0	0	0	90	0	0	0	0	0

From physical optics, it can be shown that the optimum width of a concave grating is given by

$$W_{opt} = 2.51 \left[R^3 \lambda \frac{\cos \alpha \cos \beta}{\sin^2 \alpha \cos \beta + \sin^2 \beta \cos \alpha} \right]^{\frac{1}{4}}. \qquad (2.24)$$

This is essentially the same result as derived from geometrical optics with the exception that the numerical constant is then equal to 2.38. Values of W_{opt} are given in Table 2.2 as a function of wavelength and angle of incidence for a 1 m grating with $m/d = 6000$ cm^{-1} and 12000 cm^{-1}. To obtain the value of W_{opt} for any radius R cm, multiply the values in Table 2.2 by $(R/100)^{\frac{3}{4}}$.

Because W_{opt} decreases when either α or β, or both, increase, then the resolving power will also decrease. Thus it is always desirable to work with the *inside* spectrum in order to minimize β. For example although the dispersion tends to infinity as $\beta \to \pi/2$, $\mathcal{R} \to 0$. For high resolution it is thus not desirable to use greater angles of incidence and diffraction than are necessary for sufficient reflection.

2.4 ASTIGMATISM

The major aberration of a concave mirror is astigmatism, and this imperfection is inherited by the concave diffraction grating. The theory of the astigmatism of a concave grating was first developed by Runge and Mannkopf [10]. More recently it has been dealt with in detail by Beutler [7] and Namioka [8].

Astigmatism results in a point on the slit being imaged into a vertical line; that is, focusing is achieved only in the horizontal plane. The length of the astigmatic image z is given by [7,8]

$$z = \left[l \frac{\cos \beta}{\cos \alpha} \right] + L[\sin^2 \beta + \sin \alpha \tan \alpha \cos \beta], \qquad (2.25)$$

where the first term gives the contribution due to an object slit (or entrance slit) of finite vertical length l, and the second term is the astigmatism produced by a point on the object slit. L represents the length of the ruled lines illuminated. As can be seen from eq. (2.25), the image becomes less astigmatic for near normal incidence and quite stigmatic for $\alpha = \beta = 0°$. Astigmatism is most severe at grazing angles. Figure 2.5 gives the lengths of the astimatic images of a point source on the slit in units of length of the rulings for a 600 line per mm grating. Thus for an angle of incidence $\alpha = 88°$ with rulings 2 cm long, the astimatic image of a point source at 500 Å is 16 cm long.

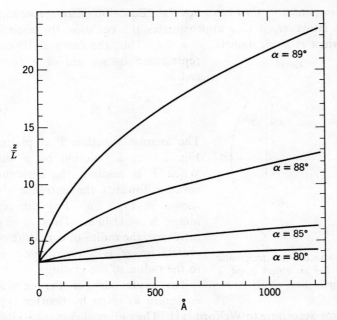

Fig. 2.5 Length of the astigmatic image z divided by the length L of the rulings illuminated for a 600 line per mm grating as a function of wavelength and angle of incidence α (courtesy T. Namioka [8]).

In general, the astimatic image is not a straight line. The curvature of the images has been studied by Beutler [7], who identified two types of curvature. The curved spectral lines caused by the astigmatism of a point source at the entrance slit was called the *astigmatic curvature*. Another kind of curvature is caused by the finite length of the entrance slit when illuminated, and was called by Beutler the *enveloping curvature*. Referring to Fig. 2.6, let the point A on the Rowland circle be focused as the astigmatic image BCD. The length of the chord z is given by (2.25). The sagittal distance x as calculated by Beutler for the case of astigmatic curvature is given by

$$x = \frac{z^2}{8R}\Psi, \qquad (2.26)$$

where R is the diameter of the Rowland circle and Ψ is the angular function

$$\Psi = \frac{1}{\Gamma^2}\left[\sin\alpha\tan^2\alpha - \sin\beta + \frac{\tan\beta}{\cos\beta}(1-\Gamma)^2\right], \qquad (2.27)$$

where $\Gamma = (\sin^2 \beta + \sin \alpha \tan \alpha \cos \beta)$. Equation 2.26 represents a parabola. However, it also approximates the equation for a circle of radius r when $r^2 \gg x^2$, namely, $z^2 = 8rx$. Thus the curve BCD can be represented by an arc of a circle of radius

$$r = \frac{R}{\Psi}. \qquad (2.28)$$

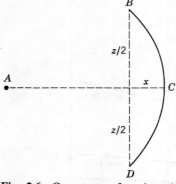

Fig. 2.6 Curvature of astigmatic image BCD for an object point A on the entrance slit.

The angular function Ψ is plotted in Fig. 2.7 as a function of α and β. When Ψ is positive, the curvature is concave towards the entrance slit as shown in Fig. 2.6. For the central image, $\Psi = 1/\sin \alpha$. Thus at grazing incidence, the radius of curvature of the central image is approximately equal to the radius of the grating.

The expression for the enveloping curvature as given by Beutler appears to be in error according to Welford [11]. The radius of curvature given by Welford is

$$r_e = \frac{R}{\Phi}, \qquad (2.29)$$

where

$$\Phi = \frac{(\sin \alpha + \sin \beta)}{\cos^2 \beta}. \qquad (2.30)$$

For the central image, the curvature $(1/r_e)$ is zero. The function Φ is plotted in Fig. 2.8. From Figs. 2.7 and 2.8, it can be seen that for the inside spectra, the astigmatic curvature is generally greater than the enveloping curvature in the vacuum uv. Although the astigmatic curvature is greatest near normal incidence, its effect is not important since the height of the astigmatic image is nearly zero. The enveloping curvature is very small near normal incidence thus the spectral lines are effectively straight. As the angle of incidence increases, the length of the astigmatic image increases more rapidly than the decrease in the astigmatic curvature; that is, the sagittal distance x increases with the angle of incidence. The curvature of the image lines limits the wavelength resolution obtainable by a monochromator that uses straight parallel exit slits. For example, for a straight exit slit of 1 cm in length in a Seya-monochromator ($\alpha = 30°15'$), the value of x for the central image is 22 μ, considering the astigmatic curvature only. When the exit slit width is very much greater than x, the

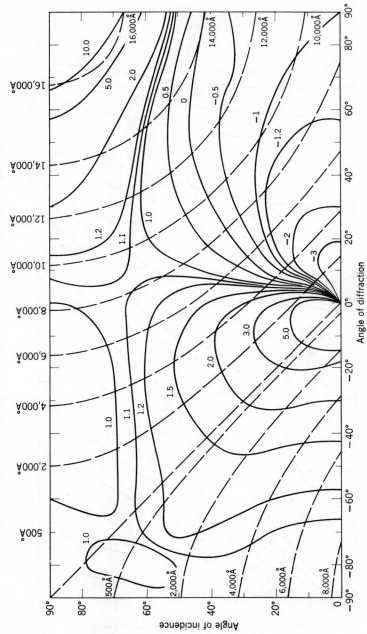

Fig. 2.7 The angular function ψ plotted for all angles of incidence and diffraction (courtesy H. G. Beutler [7]).

17

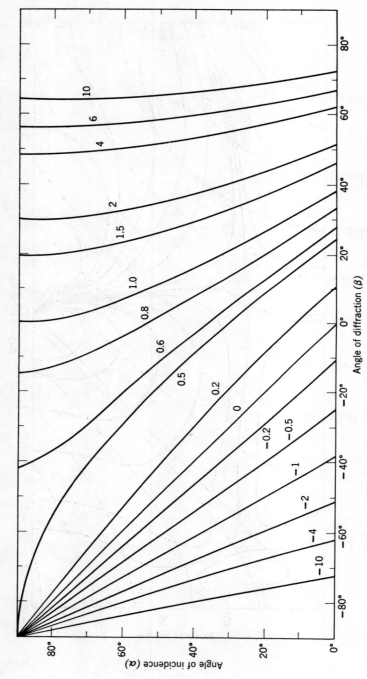

Fig. 2.8 The angular function Φ plotted for all angles of incidence and diffraction.

18

resolution impairment due to the curvature of the spectral lines is negligible and only becomes important when x and the slit width are of the same order of magnitude.

Astigmatism can be tolerated in spectroscopy since only horizontal focusing is required to separate the various wavelengths. However, it reduces the light intensity per unit area of the image and imposes strict focusing conditions to produce maximum resolution. Techniques for the reduction or elimination of astigmatism are discussed below.

2.5 STIGMATIC IMAGES

Under certain conditions it is possible to produce vertical focusing with a concave grating; that is, horizontal lines will be imaged as horizontal lines. This geometrical relation between object and image was first pointed out by Sirks [12]. He noted that a point source of light placed at N, where the normal to the grating meets the Rowland circle (Fig. 2.9), would be brought to a horizontal focus at S and to a vertical focus along ABC, an extension of the tangent to the Rowland circle at N. Thus a horizontal

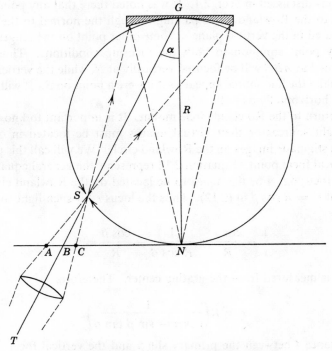

Fig. 2.9 Sirks' stigmatic focusing condition.

line of narrow rectangular aperture placed along *ABC* will come to a vertical focus at *N*. The stigmatic conditions hold strictly for rays diffracted along the normal and approximately for a short distance on either side of the normal. These conditions led Sirks to propose the following geometrical arrangement for the superposition of a comparison spectrum, a technique previously suitable only with stigmatic spectrographs. Placing a short but rather broad prism (2 or 3 mm in height) along *ABC*, he reflected sunlight through the slit *S* into the spectrograph, which gave a narrow solar spectrum with perfectly defined edges passing through the center of the field. At the same time, light from a sodium flame placed at *T* passed over the top and bottom of the prism to form an astigmatic spectrum above and below the solar spectrum. This arrangement covers a very small spectral region at one setting, and, since the spectrum must be viewed along the normal to the grating, *S* must be moved around the Rowland circle to observe another spectral region. The secondary focal plane *ABC* is always positioned behind *S* at a distance

$$SB = R \sin \alpha \tan \alpha. \tag{2.31}$$

Sirks' focusing condition is a special case of the vertical focusing conditions discussed in Sect. 2.1. It was noted there that any point on the tangent to the Rowland circle passing through the normal to the grating was focused in the vertical plane at some other point on the tangent. *N* is the only point common to both the focusing conditions. Thus a line illuminated at *ABC* will be focused vertically at *N*, while the vertical slit at *S* will make the line source appear as if it were a point on *S*. It will thus be focused horizontally as well.

We return to the Rowland circle mount. It is important to know where point light sources or their virtual images must be located in order to produce stigmatic images on the Rowland circle. We will call this position the vertical focal point. Equation 2.12 represents the general equation for vertical focusing. For the image to be located on the Rowland circle, we must put $r' = R \cos \beta$ in (2.12). Thus the locus r for such light sources is given by

$$\frac{1}{r} - \frac{\cos \alpha}{R} + \frac{1}{R \cos \beta} - \frac{\cos \beta}{R} = 0,$$

where r is measured from the grating center. Therefore,

$$r = R\left(\frac{1}{\cos \alpha - \sin \beta \tan \beta}\right). \tag{2.32}$$

The distance l between the primary slit *S* and the vertical focal point is given by $l = r - r_0$, where r_0 is the distance between *S* and the grating and is equal to $R \cos \alpha$; that is,

$$l = R\left(\frac{1}{\cos \alpha - \sin \beta \tan \beta} - \cos \alpha\right). \qquad (2.33)$$

For $\beta = 0$, (2.33) reduces to $l = R \sin \alpha \tan \alpha$, which is identical to (2.31).

The fact that a stigmatic image for a given wavelength can be achieved using a concave diffraction grating has increased the usefulness of the concave grating. An important application of (2.33) has been in solar spectroscopy using rocket-borne spectrographs [13–16]. Since the light source (the sun in this case) is at infinity, then in (2.33), $l = \infty$. This will be true when

$$\cos \alpha = \sin \beta \tan \beta.$$

From the grating equation, we have the further condition that

$$\frac{m\lambda}{d} = \sin \alpha + \sin \beta.$$

For a given order number, wavelength, and grating spacing these two equations can be solved simultaneously for α and β. Consider the Lyman-α line at 1215.7 Å imaged in the first order by a 600 line per mm grating. Then to produce a stigmatic image of Lyman-α on the Rowland circle with parallel incident light, we find

$$\alpha = 49.5° \qquad \text{and} \qquad \beta = -43.4°.$$

In the more general case a concave mirror may be used to focus the object onto the entrance slit of the spectrograph [14,16]. This has the added advantage of increasing the light-gathering power of the instrument. Figure 2.10 shows a typical arrangement for the removal of astigmatism at some wavelength for a grazing incidence mount. The mirror must provide a horizontally focused image at S_2, the entrance slit of the spectrograph, and a vertically focused image at L, the vertical focal point for the grating. The distance GL is given by (2.32). As can be seen from (2.32), when α and β are large, r is negative, and the point L is a virtual focal point. However, the above arrangement to eliminate astigmatism applies whether r is negative or positive. A toroidal mirror can be used instead of a spherical mirror giving more freedom in the choice of the magnitude of the optical parameters. The radii of the toroidal mirror and its position relative to the spectrograph are given by the following relations:

HORIZONTAL FOCUS $\qquad \dfrac{1}{s} + \dfrac{1}{s_h} = \dfrac{2}{r_h \cos \varphi}, \qquad (2.34)$

VERTICAL FOCUS $\qquad \dfrac{1}{s} + \dfrac{1}{s_v} = \dfrac{2 \cos \varphi}{r_v}, \qquad (2.35)$

where r_h and r_v are the radii in the plane and at right angles to the plane of the Rowland circle, respectively. The distance s denotes the object distance of the aperture or slit S_1, and φ is the angle of incidence. The combination of a toroidal mirror and a spherical concave grating provides the best method for reducing astigmatism over a wide wavelength range for a grazing incidence spectrograph. For other angles of incidence, typically the Seya-Namioka mount, an ellipsoidal or toroidal grating is better. This is discussed further in the next section.

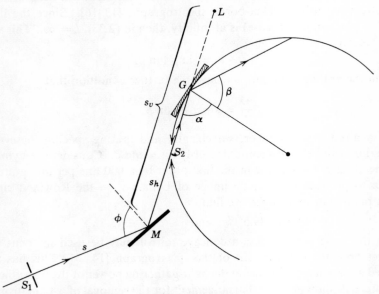

Fig. 2.10 Removal of astigmatism at a given wavelength using a foregrating or a concave mirror M. The main spectrograph is defined by the entrance slit S_2 and the grating G. L is the common vertical focal point for both M and G. $S_1M = s$; $MS_2 = s_h$; $ML = s_v$; $S_2G = r_0$; $S_2L = l$; $GL = r$.

When working in the higher orders of a grating spectrograph, it is desirable to eliminate overlapping orders. In the vacuum ultraviolet region this can be achieved by using a foregrating [17,17a]. When a spherical or toroidal foregrating is used in place of the mirror in the above discussion, astigmatism is removed at one wavelength and reduced in the vicinity of that wavelength. It should be noted that the foregrating system consisting of slits S_1 and S_2 and the foregrating all lie on a Rowland circle of diameter equal to the radius of the foregrating (see Fig. 2.10). S_2 is common to both grating systems. Equation 2.33 is applied first to the main spectrograph to locate its vertical focal point. Then it is applied to the foregrating system so that S_1 is imaged vertically at the vertical focal point of the main spectrograph. That is, the position L in Fig. 2.10 is the vertical focal point for both the foregrating and the main grating system. The remaining parameters of the foregrating are determined from the grating equation $m\lambda = d(\sin \alpha + \sin \beta)$.

2.6 ELLIPSOIDAL AND TOROIDAL GRATINGS

Although no ellipsoidal gratings exist at present, their properties are promising and will be discussed briefly.

The theory of the ellipsoidal diffraction grating has been developed by Namioka [18]. The grating equations are identical to (2.15) and (2.16) for the spherical grating. The focal condition is given by

$$\frac{\cos^2 \alpha}{r} - \frac{a}{b^2} \cos \alpha + \frac{\cos^2 \beta}{r'} - \frac{a}{b^2} \cos \beta = 0, \qquad (2.36)$$

where a, b, and c are the lengths of the semi-axes of the ellipsoid in the directions of the x, y, and z axes, respectively. The other symbols and the geometrical arrangement are as shown in Fig. 2.2 for the spherical grating. A solution of (2.36) is

$$r = \frac{b^2}{a} \cos \alpha \quad \text{and} \quad r' = \frac{b^2}{a} \cos \beta. \qquad (2.37)$$

Equation 2.37 represents a circle of diameter b^2/a in the xy plane similar to the Rowland circle.

The stigmatic condition for an ellipsoidal grating is given by

$$\frac{c^2}{b^2} = \cos \alpha \cos \beta, \qquad (2.38)$$

which is shown graphically in Fig. 2.11. When (2.38) does not hold, the length of the astigmatic image of a point source is given by

$$z = L(1 + \sec \alpha \cos \beta) \left| 1 - \frac{b^2}{c^2} \cos \alpha \cos \beta \right|. \qquad (2.39)$$

The optimum grating width is

$$W_{\text{opt}} = 2.12 \left[\frac{2R^3 d}{m} \tan \left(\frac{\alpha + \beta}{2} \right) \frac{\cos \alpha \cos \beta}{|1 - (R/a) \cos \alpha \cos \beta|} \right]^{1/4}, \qquad (2.40)$$

where $R = b^2/a$ is the diameter of the equivalent Rowland circle.

In the region where

$$\cos \beta \leq \frac{2c^2}{b^2 + c^2} \sec \alpha \qquad (2.41)$$

is valid, and when $a = b \neq c$ or $a = c \neq b$, the ellipsoidal grating gives less astigmatism than the spherical grating. However, for $b = c$, nothing is gained.

In theory the ellipsoidal grating has a better optical performance than the spherical grating both at grazing incidence and in the Seya–Namioka

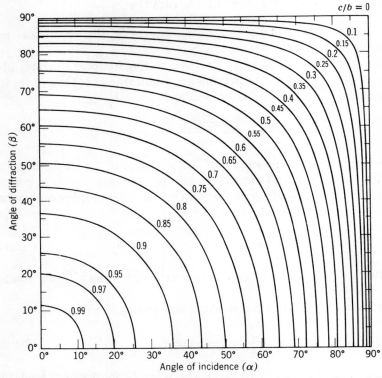

Fig. 2.11   The function $c/b = (\cos \alpha \cos \beta)^{\frac{1}{2}}$ plotted as a function of α and β. This is the stigmatic condition for an ellipsoidal grating. b and c are the lengths of the semi-axes of the ellipsoid in the direction of the z and y axis, respectively, both measured in the xy plane (courtesy T. Namioka [18]).

mounting. However, it is less efficient in reducing astigmatism at grazing incidence than the combination of toroidal mirror and spherical grating.

The reduction of astigmatism has been considered theoretically by several authors. Haber [19] has presented the theory of the toroidal grating, showing that it is free from astigmatism at two wavelengths with practically no astigmatism in the vicinity of those wavelengths. Greiner and Schäffer [20] have discussed astigmatism caused by both spherical and toroidal gratings, while Sakayanagi [21] has considered the reduction in astigmatism using a spherical concave grating ruled with circular grooves.

Schönheit [21a] has successfully produced and used a toroidal grating for the reduction of astigmatism in a Seya–Namioka monochromator. The radii of the grating can be determined from the equations for horizontal and vertical focusing, namely,

$$\frac{\cos^2 \alpha}{r} + \frac{\cos^2 \beta}{r'} - \frac{\cos \alpha}{R_h} - \frac{\cos \beta}{R_h} = 0 \qquad (2.9)$$

and

$$\frac{1}{r} + \frac{1}{r'} - \frac{\cos \alpha}{R_v} - \frac{\cos \beta}{R_v} = 0, \qquad (2.12)$$

where R_h and R_v are the radii of the grating in the plane and at right angles to the plane of the Rowland circle, respectively. For the central image, α is equal to β and $r = r'$. Substituting these values into (2.9) and (2.12), we obtain

$$\frac{R_v}{R_h} = \cos^2 \alpha_0,$$

where α_0 is the angle of incidence when viewing the central image. In the Seya mount, $\alpha_0 \sim 35°$, hence $R_v/R_h \sim 0.671$, which is the ratio used by Schönheit. For small rotations of the grating, the images remain approximately stigmatic.

2.7 GRATING EFFICIENCY

The efficiency of a grating at a given wavelength can be defined as the percentage of the incident radiant flux returned by the grating into a given spectral order.

The groove separation of a diffraction grating influences the angular dispersion of the diffracted radiation. The groove shape, on the other hand, controls the amount of radiation concentrated into a given order. The concentration of radiation into any desired spectrum was first discussed by Lord Rayleigh in 1888. However, it was not until 1910 that the first successful grating was ruled by R. W. Wood [22] with grooves of controlled shape. These gratings were called *echelette* gratings. Because it is now standard practice to rule gratings with grooves of controlled shape, the term *echelette* has been dropped. In its place we sometimes refer to a grating as being blazed at a given wavelength. A discussion of the echelle grating is given in Chapter 3.

Figure 2.12 shows a cross section of a typical blazed grating. Let N be the normal to the macroscopic surface of the grating while N' denotes the normal to the facets. The grooves are cut such that the facets make an angle θ with the grating surface. The angles α and β are, as before, the

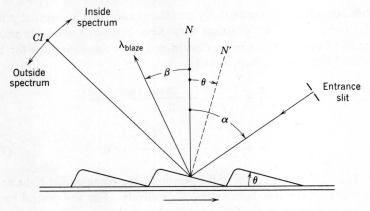

Fig. 2.12 Cross section of a blazed grating. θ is the blaze angle.

angles of incidence and diffraction. The principle of concentrating radiant energy into a given wavelength is that this wavelength must be diffracted in a direction which coincides with the direction of the specularly reflected beam from the surface of the facet. Referring to Fig. 2.12, this condition is expressed by

$$\alpha - \theta = \beta + \theta,$$

hence

$$\theta = \frac{\alpha - \beta}{2}.$$ (2.42)

However, for Fig. 2.12,

$$m\lambda = d(\sin \alpha - \sin \beta).$$ (2.43)

Eliminating β from (2.42) and (2.43), we obtain the wavelength λ_{blaze} for which the grating is blazed, namely,

$$m\lambda_{\text{blaze}} = 2d \sin \theta \cos (\alpha - \theta).$$ (2.44)

For normal incidence, $\alpha = 0$, and (2.44) becomes

$$m\lambda_{\text{blaze}} = d \sin 2\theta.$$ (2.45)

Equation 2.44 is quite general, provided the sign of θ is taken to be positive when it is on the same side of the normal as α, and negative if it is on the opposite side. It can be seen from (2.44) or (2.45) that a grating blazed for 2000 Å in the first order ($m = 1$) is also blazed for 1000 Å in the second order and for 500 Å in the fourth order, etc. Under the conditions of illumination shown in Fig. 2.12, the concentration of radiation can only occur in the inside spectrum. Should the radiation be incident on the opposite side of the normal N or, equivalently, if the grating is rotated through 180°, θ will be negative and the blaze will occur at the same wavelength, but in the outside spectrum. When the grating is illuminated in this direction, however, a fraction of the diffracted beam will always strike the steep edge of the grooves thus producing some scattered radiation and diminishing the intensity of the diffracted beam. The grating is

Fig. 2.13 Electron micrograph of test rulings in aluminum. Note that the ruled area is smoother than the unruled portion (courtesy D. Richardson and C. F. Mooney, Bausch and Lomb Co.).

most efficiently used with the radiation incident "uphill," as shown in Fig. 2.12. Bausch and Lomb place an arrow on the back of their gratings that points towards the blaze angle. Therefore, the arrow should point towards the incoming radiation. Figure 2.13 shows an electron micrograph of test rulings made by the Bausch and Lomb Company on an aluminized mirror. The shadow cast by a metal vapor incident obliquely on an asbestos fiber reveals the shape of the grooves. The white areas of the photograph indicate the shadow of the asbestos rod. Anderson et al.

[23] have described the Bausch and Lomb shadow-casting technique and the method for analyzing the groove shape from electron micrographs. Figure 2.14*b* is a schematic of the photograph shown in Fig. 2.13 and indicates the dimensions to be used in determining the blaze angle θ. Figure 2.14*a* shows a section through the fiber and grating cut parallel to the rulings; the fiber rod is assumed to be at right angles to the rulings and

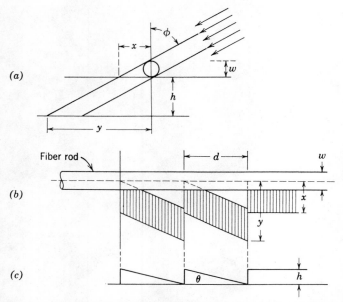

Fig. 2.14 Determination of blaze angle from electron micrograph. (*a*) Section through the fiber and grating cut parallel to the rulings. The arrows indicate the direction of the shadow-casting metal vapor, which strikes the top of the groove at a distance *x* and the bottom of the groove at a distance *y* from the center of the fiber rod. (*b*) Plan view of grating surface showing the shadow cast by the fiber rod. (*c*) Section cut through grating at right angles to the rulings.

of uniform diameter *w*. The direction of the metal vapor is shown by the oblique arrows incident at an angle φ to the normal of the grating surface. If this particular section is taken through the bottom of the groove, then the height *h* represents the height of the grooves shown in Fig. 2.14*c*. The metal vapor strikes the top of the groove at a distance *x* from the center of the fiber and strikes the bottom of the groove at a distance *y* from the center of the fiber. From Fig. 2.14*a*, the height of the grooves is given by

$$h = w\left(\frac{y}{x} - 1\right)\left(1 + \frac{1 - \sin \varphi}{2 \sin \varphi}\right). \tag{2.46}$$

The blaze angle can then be determined from $\tan \theta = h/d$. The term $(1 - \sin \varphi)/(2 \sin \varphi)$ can be neglected when φ is greater than about 65°. The angle of shadow-casting φ can be determined approximately from the relation $\tan \varphi \approx x/w$. To determine the angle φ more precisely, Anderson et al. placed a latex sphere on the surface of the grating before shadow casting. Then from the ratio of the length to the width of the shadow of the sphere, they were able to determine $\tan \varphi$. Analyzing Fig. 2.13, we find that the blaze angle is about 4.5°, the step height approximately 1300 Å, and the angle of shadow-casting about 77°. The scale of the photograph can be determined from the known groove separation d, which in this case is equal to 1.67 μ (600 grooves per mm). The electron micrographs show that grooves can be ruled to conform with the ideal shape shown in Fig. 2.12. It is very revealing to see the surface of the ruled and unruled areas under the high magnification produced by the electron microscope. As can be seen from Fig. 2.13, the unruled area of the evaporated aluminum coating is quite rough, whereas the ruled area, especially at the bottom of the groove, is considerably smoother. Replica gratings made from such masters often show superior performance since the rough edges at the top of the grooves in the master are at the bottom of the grooves in the replica.

For further discussions on the efficiencies and production of diffraction gratings, the reader is referred to references 24 to 31.

In some investigations, it is necessary to know the actual efficiency of a grating as a function of wavelength. This knowledge is necessary, for example, in determining the spectral energy distribution from sources such as the sun or high temperature plasmas. The measurement of grating efficiencies can be readily achieved using the experimental arrangement shown in Fig. 2.15 [32]. Other arrangements have been described in the literature using polarized and unpolarized incident radiation [33–35]. In this particular arrangement, radiation from the exit slit of a monochromator enters the test chamber and is baffled in order to illuminate a small portion of the grating surface. By changing the position of the baffle, the efficiencies of different areas of the ruled surface can be studied.

To obtain the amplitude of the various diffracted orders, the photomultiplier is rotated about the grating. The output current and position of the photomultiplier relative to the incident beam is recorded when the output current reaches a maximum value. The grating is then moved out of the path of the light beam by raising it vertically to allow the irradiance of the incident beam to be measured. The ratios of the output currents due to reflected orders to that due to the incident beam gives the required efficiencies.

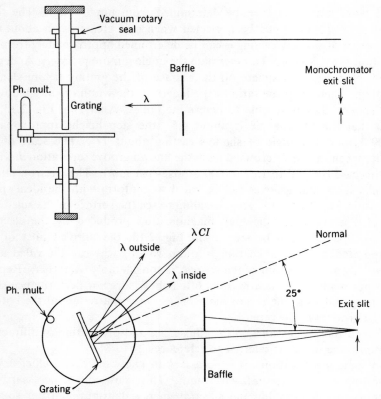

Fig. 2.15 Experimental arrangement for measuring grating efficiencies (Samson [32]).

Figures 2.16 and 2.17 illustrate typical diffraction patterns found by rotating the photomultiplier about the grating. Figure 2.16 was obtained using a 600 line per mm grating while a 1200 line per mm grating was used in Fig. 2.17. The abscissae are expressed in degrees zeroed on the central image, labelled *CI*, whereas the ordinate axes represent the percent of the incident radiant flux which is reflected from the grating into the various orders. Figure 2.17 also shows the development of a blaze at 686 Å for the inside first order.

Figure 2.18 shows the efficiencies of the inside first order for three areas of the 1200 line per mm grating as a function of wavelength. Figure 2.19 is a plot of the ratio of the inside first order intensities to that of the sum of all orders, that is, to the total reflectance, for the 1200 line per mm grating. The maxima of such curves represent the wavelength at which the grating is blazed. The curves labelled 1, 2, and 3 refer to the three portions of the grating studied. The dashed curve is the ratio of the inside second order

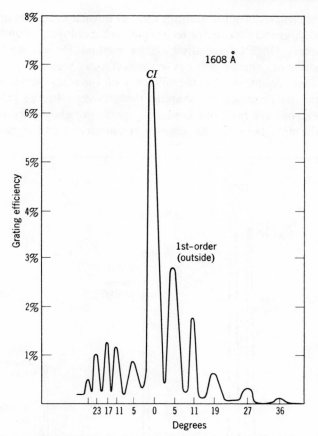

Fig. 2.16 Monochromatic radiation of wavelength 1608 Å diffracted by a particular grating ruled with 600 lines per mm (Samson [32]).

to that of the total reflectance. The maximum of this curve indicates a blaze at 650 Å, approximately half the wavelength at the maximum of the solid curve 3. From Fig. 2.19 it can be seen that the blazed wavelength shifts towards shorter wavelengths as the area of the grating illuminated changes from area 3 toward area 1. Thus, the ruling of this grating apparently produced a continuously changing blaze angle. To maintain a more constant blaze angle over the entire surface of a grating it should be ruled in three portions, the angle of the ruling diamond being reset after each ruling. Such *tripartite* gratings are available commercially. The resolution of such gratings, however, cannot be greater than that of a single ruled portion.

This discussion is applicable to the study of grating efficiencies at any angle of incidence. However, for radiation shorter than approximately

300 Å the efficiencies of most gratings used at normal incidence are poor and, generally, grazing incidence techniques are employed. Tomboulian and co-workers [36,37] have discussed a method for comparing the reflecting power of concave gratings in the soft x-ray region.

There is some doubt whether the efficiency of a grating used at grazing incidence can be increased by shaping the grooves. In the past, best results appear to have been obtained using gratings lightly ruled on glass such that the area between the grooves is untouched and is extremely

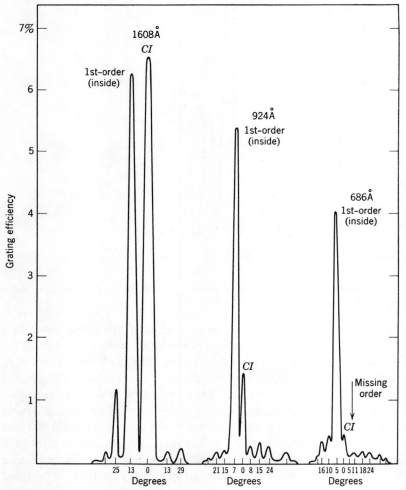

Fig. 2.17 Monochromatic radiation of wavelengths 1608, 924, and 686 Å diffracted by a particular grating ruled with 1200 lines per mm. The development of a blaze in the first-order is followed through from 1608 Å to its maximum at 686 Å (Samson [32]).

smooth [25,26]. The principle of this argument is that the efficiency of a grating will increase when the radiation is incident at a sufficiently large angle of incidence to experience total reflection from the surface. In this case the reflecting surface must be very smooth. The principle of a controlled groove shape is still valid at these extreme angles and would be expected to improve the grating efficiency. However, the problem may be to

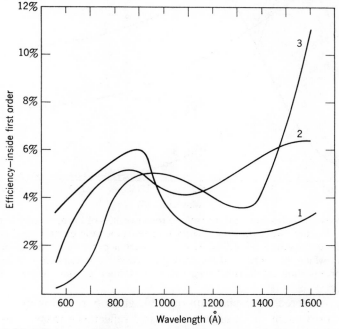

Fig. 2.18 Efficiency of the inside first-order spectrum measured for three areas of the grating as a function of wavelength (1200 lines per mm) (Samson [32]).

rule the grating such that the reflecting facets have the required degree of smoothness. If this can be achieved, a blazed grating might be expected to be superior at grazing incidence angles. Gabriel et al. [38] have, in fact, obtained a factor of 2 in efficiency using a platinized replica grating with a blaze angle of 1.5° compared to a lightly ruled glass original grating. They used an angle of incidence of 88° with the radiation incident "uphill" as shown in Fig. 2.12. At this angle of incidence, the grating will be blazed for 53 Å. It is unclear, however, whether the increased efficiency was due to the blazed grating or the high reflectance of the platinum overcoating.

As mentioned above, the efficiency of a grating is also influenced by the reflectance of the material in which the grating is ruled or has been coated. Watanabe [39], using a platinized capillary in a hydrogen dc light source,

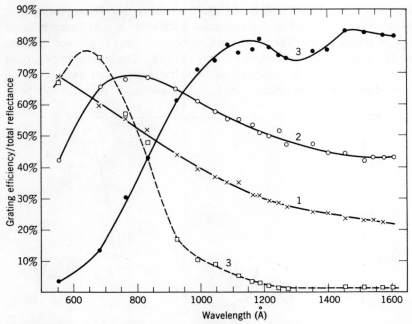

Fig. 2.19 The solid curves represent the ratio of the intensities of the first-order to that of the sum of all orders, while the broken curve represents the ratio of the second-order intensity to that of the sum of all orders. Hence, the maxima of the curves represent the wavelengths at which that portion of the grating is blazed. The numbers 1, 2, and 3 refer to the three portions of the ruled area studied (1200 lines per mm) (Samson [32]).

accidentally sputtered some platinum onto his grating, resulting in a considerable improvement in its efficiency. Early reflectance measurements on several materials by Sabine [40] had shown the superiority of platinum over other materials for wavelengths shorter than 1000 Å.

   Recently there has been a considerable amount of work published on the reflectance of materials and efficiencies of gratings with different coatings.

2.8   REFLECTIVE COATINGS

   From the elementary theory of metals, the critical angle of reflection for radiation of wavelength λ is given by

$$\sin \theta_c = \lambda \left(\frac{e^2}{mc^2} \frac{N}{\pi} \right)^{1/2}, \tag{2.47}$$

where θ_c is the grazing angle of incidence, and N is the number of electrons per unit volume. Thus, for a given wavelength, total reflection occurs at all grazing angles up to a certain maximum θ_c at which point the reflectance drops rapidly. Conversely, for a given grazing angle of incidence, λ in (2.47) represents the minimum wavelength which is reflected, all wavelengths longer than this minimum being totally reflected. Inserting the values of the constants in (2.47), we obtain

$$\lambda_{min}(\text{Å}) = (3.33 \times 10^{14})N^{-\frac{1}{2}} \sin \theta. \tag{2.48}$$

For a glass or aluminized grating, $N^{\frac{1}{2}} = 8.8 \times 10^{11}$, therefore for small θ,

$$\lambda_{min}(\text{Å}) = 6.6 \, \theta°. \tag{2.49}$$

Thus for a grazing angle of incidence of 1°, the minimum reflected wavelength is 6.6 Å. Of course, the theory does not hold exactly. Nevertheless, it can be used to give an order of magnitude estimate. Table 2.3 lists the

Table 2.3 The Minimum Wavelength Reflected for a Given Grazing Angle of Incidence for Several Grating Materials and Impurities

Material	Density	Number of Electrons per cc N	$N^{\frac{1}{2}}$	$\lambda_{min}(\text{Å})$
Pentadecane (oil)	0.77	27×10^{22}	5.2×10^{11}	$641 \sin \theta$
Glass	2.6	78×10^{22}	8.8×10^{11}	$379 \sin \theta$
Aluminum	2.7	78×10^{22}	8.8×10^{11}	$379 \sin \theta$
Aluminum oxide	3.9	115×10^{22}	10.7×10^{11}	$312 \sin \theta$
Silver	10.5	276×10^{22}	16.6×10^{11}	$201 \sin \theta$
Gold	19.3	466×10^{22}	21.6×10^{11}	$154 \sin \theta$
Platinum	21.4	514×10^{22}	22.7×10^{11}	$147 \sin \theta$
Iridium	22.4	542×10^{22}	23.3×10^{11}	$143 \sin \theta$

value of λ_{min} as a function of the grazing angle θ for several materials used for overcoating gratings along with two common impurities which are often present on the surface of gratings, namely, oil and oxides [41]. It can be seen from the table that as the density of the reflecting surface increases, the shorter is the wavelength which can be reflected. Platinum and iridium have about the highest density of all the elements and should, therefore, reflect the shortest wavelength for a given angle θ. The efficiency of platinum has been demonstrated by Gabriel et al. [38] who recorded a minimum wavelength of 4.7 Å using a grazing angle of 2°. From Table 2.3, we see that the theoretical value of λ_{min} for platinum at this angle of incidence is 5.3 Å. Gold is not quite so dense as platinum, but it is much easier to evaporate. Oil, on the other hand, has a very small density. Therefore, should a gold or platinum grating accumulate a very thin

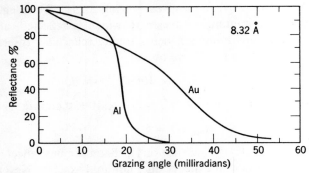

Fig. 2.20   Reflectance of Al and Au at 8.32 Å as a function of the grating angle (courtesy R. W. Hendrick [42]).

coating of oil, it would seriously affect the reflectance of the grating near its short wavelength reflectance limit. This effect has been observed in practice.

The reflectances of several elements have been measured by Hendrick [42] at 8.32 Å as a function of the grazing angle of incidence. His results for Al and Au are shown in Fig. 2.20. The critical angle of reflectance is generally taken to be the angle at which the slope of the reflectance curve is a maximum. For Al, Hendrick measured a critical angle of 18.7 milliradians, and for Au, a value of 37.2 milliradians. Similar measurements have been made by Johnson and Wuerker [43] at 44.6 and 114 Å. Their results for gold are shown in Fig. 2.21 and agree very well with those of Lukirskii, Savinov, and Shepelev [44]. Since gold has many absorption edges in the wavelength range 1 to 175 Å, it is possible that observed spectra may show false structure in the vicinity of these edges. However,

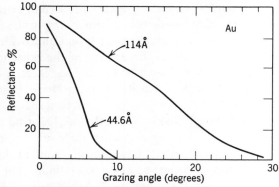

Fig. 2.21   Reflectance of Au at 114 and 44.6 Å as a function of the grazing angle of incidence (courtesy G. L. Johnson and R. F. Wuerker (43)).

a study by Lukirskii, Zimkina, and Brytov [45] showed that disturbances in this wavelength range were not serious when the angle of grazing incidence was 5.5°. As the grazing angle increases, the effect of absorption edges on the reflectance becomes more pronounced [46]. The complex index of refraction of many elements and compounds has been determined by Lukirskii et al. [47] from reflectance measurements at wavelengths between 23.6 and 113 Å measured as a function of the angle of incidence.

For radiation shorter than about 1000 Å, platinum is generally the most useful and efficient reflector, especially at normal incidence. Unlike

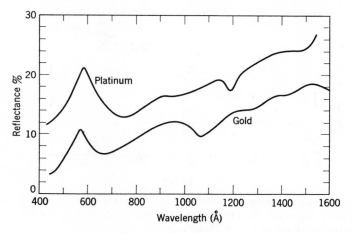

Fig. 2.22 Normal incidence reflectance of Au and Pt as a function of wavelength.

aluminum, platinum shows very little change in reflectance after exposure to air [48,49]. However, the actual magnitude of the reflectance of most materials including platinum depends critically on the method of producing the reflecting surface. Most reflecting films are produced by vacuum evaporation either from resistance-heated filaments or by induction heating. For highest quality reflecting films, it is generally necessary to deposit the films as rapidly as possible and at a high vacuum, typically 5×10^{-6} torr or better. To obtain a smooth surface for maximum reflectance, the deposited film is usually kept as thin as possible while still being opaque to the incident radiation. For platinum, the optimum thickness is about 100 Å as measured by Jacobus et al. [49]. The reflectance was down about 10 percent for a 500 Å thick film.

Gold appears to be a second choice for a suitable reflector for wavelengths shorter than 1000 Å. Like platinum, gold is very stable on exposure to air and provides the highest reflectance for film thicknesses of approximately 100 Å [50]. Figure 2.22 shows the reflectance of gold and platinum between 400 and 1600 Å. The increase in reflectance around 584 Å appears to be characteristic of both gold and platinum. The results for gold shown in Fig. 2.22 agree well with those of Canfield et al. [50] (see Chapter 9). The published data on platinum [49,51,52] tend to lie both above and below those shown in Fig. 2.22, emphasizing the variation which can be obtained under the different conditions of production.

In the past aluminum has been the most commonly used material to coat mirrors and gratings. However, Hass et al. [53] have shown that although aluminum is an excellent material for use above 2000 Å, it has a

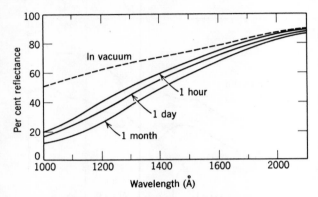

Fig. 2.23 Normal incidence reflectance of best quality aluminum films before and after 1 hour, 1 day, and 1 month exposure to air as a function of wavelength (courtesy P. H. Berning, G. Hass, and R. P. Madden [54]).

very serious aging effect at shorter wavelengths. This aging effect has been traced to the formation of an oxide layer. As the thickness of the aluminum oxide layer builds up, the reflectance continually decreases. When freshly deposited aluminum is irradiated with ultraviolet light, there is a speed up in the oxide formation. Figure 2.23 illustrates this aging effect, showing the reflectance of aluminum measured in vacuum immediately after deposition and then remeasured after an exposure of one hour, one day, and one month to air [54]. Hass and Tousey [52] showed that the excellent reflectance of aluminum could be maintained by preventing the growth of oxide layers. To prevent oxidation, they overcoated freshly deposited aluminum with a thin protective layer of magnesium fluoride. The results were quite dramatic; the reflectance between 1200 and 2000 Å was about 80 per cent and showed no decrease on exposure to air. More recently, Canfield et al. [55] have reported results of extended tests on MgF_2 overcoatings. They have observed no aging effects nor any loss in reflectance after irradiation with 1-MeV electrons and 5-MeV protons. The optimum thickness of MgF_2 coating depends upon the wavelength. Figure 2.24

shows the reflectance of Al + MgF$_2$ as a function of the thickness of the MgF$_2$ [56]. A film thickness of approximately 220 Å is a fairly good compromise for radiation between 1025 and 1600 Å. The use of LiF overcoatings has the effect of maintaining the reflectance of aluminum at shorter wavelengths. This is because the transmission limit of LiF is slightly lower than that of MgF$_2$. The results obtained by Hunter [56] are shown

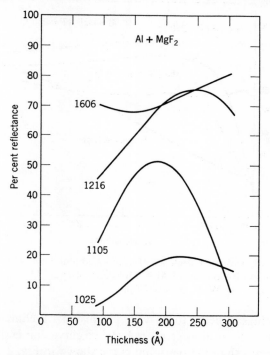

Fig. 2.24 Normal incidence reflectance of Al + MgF$_2$ as a function of the thickness of the MgF$_2$ (courtesy W. R. Hunter [56]).

in Fig. 2.25. The optimum thickness of LiF should be about 170 Å to maintain high reflectance at the shorter wavelengths. LiF, however, is rather hydroscopic and has a reduced transmittance on exposure to a humid atmosphere. Hunter prevented this by overcoating the Al + LiF combination with a very thin protective layer of MgF$_2$ about 15 Å thick.

The reflectances of many materials have been measured in the vacuum uv [57,60], and an excellent review has been given by Madden on the techniques for preparing and measuring reflective coatings [61]. At present, the most efficient reflective coatings for gratings in general use appear to be Al + MgF$_2$ for the wavelength range 2000 to 1000 Å and platinum for wavelengths below 1000 Å.

The use of a focused laser beam of high instantaneous power is currently being used by the author to produce reflective coatings of refractory materials of high density and high melting points. Thin films of a few hundred angstroms thick can be deposited in approximately 100 μsec and under conditions of ultrahigh vacuum. This technique for producing high

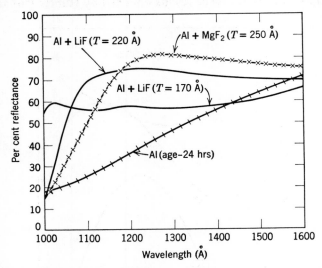

Fig. 2.25 Normal incidence reflectance of evaporated aluminum with two thicknesses of LiF overcoatings. Reflectance of Al + MgF$_2$ and 24-hour old aluminum are shown for comparison (courtesy W. R. Hunter [56]).

reflective coatings appears to be very promising. Preliminary measurements on thin iridium films give a reflectance curve that is slightly greater at all wavelengths than the platinum curve shown in Fig. 2.22.

REFERENCES

[1] H. A. Rowland, *Phil. Mag.* **13,** 469 (1882).

[2] J. S. Ames, in H. A. Rowland's *Physical Papers* (John Hopkins Press, Baltimore, 1902).

[3] G. R. Harrison, *J. Opt. Soc. Am.* **39,** 413 and 522 (1949).

[4] G. R. Harrison and G. W. Stroke, *J. Opt. Soc. Am.* **45,** 112 (1955).

[5] G. R. Stroke, in *Progress in Optics*, Vol. 2, ed. E. Wolf (North-Holland, Amsterdam, 1963) p. 1.

[6] H. A. Rowland, *Phil. Mag.* **16,** 197 and 210 (1883).

[7] H. G. Beutler, *J. Opt. Soc. Am.* **35,** 311 (1945).

[8] T. Namioka, *J. Opt. Soc. Am.* **49,** 446 (1959); **51,** 4 (1961); *see also* T. Namioka, in *Space Astrophysics*, ed. W. Liller (McGraw-Hill, New York, 1961), p. 228.

[9] J. E. Mack, J. R. Stehn, and B. Edlén, *J. Opt. Soc. Am.* **22,** 245 (1932).

[10] C. R. Runge and R. Mannkopf, *Z. Physik* **45**, 13 (1927).

[11] W. T. Welford, in *Progress in Optics*, Vol. IV, ed. E. Wolf (North-Holland, Amsterdam, 1965).

[12] J. L. Sirks, *Astronomy and Astrophysics* **13**, 763 (1894).

[13] S. C. Miller, Jr., R. Mercure, and W. A. Rense, *Astrophys. J.* **124**, 580 (1956).

[14] W. A. Rense and T. Violett, *J. Opt. Soc. Am.* **49**, 139 (1959).

[15] J. D. Purcell, D. M. Packer, and R. Tousey, *Nature* **184**, 8 (1959).

[16] F. S. Johnson, H. M. Malitson, J. D. Purcell, and R. Tousey, *Astrophys. J.* **127**, 80 (1957).

[17] A. E. Douglas and G. Herzberg, *J. Opt. Soc. Am.* **47**, 625 (1957).

[17a] H. E. Blackwell, G. S. Shipp, M. Ogawa, and G. L. Weissler, *J. Opt. Soc. Am.* **56**, 665 (1966).

[18] T. Namioka, *J. Opt. Soc. Am.* **51**, 4 (1961).

[19] H. Haber, *J. Opt. Soc. Am.* **40**, 153 (1950).

[20] H. Greiner and W. Schäffer, *Optik* **16**, 288 and 350 (1959).

[21] Y. Sakayanagi, *Sci. Light* (Tokyo) **3**, 1 (1954).

[21a] E. Schönheit, *Optik* **23**, 305 (1965/66).

[22] R. W. Wood, *Phil. Mag.* **20**, 770 (1910).

[23] W. A. Anderson, G. L. Griffin, C. F. Mooney, and R. S. Wiley, *Appl. Optics* **4**, 999 (1965).

[24] J. B. Nicholson, C. F. Mooney, and G. L. Griffin, *Advances in X-Ray Analysis* **8**, 301 (1965).

[25] L. A. Sayce and A. Franks, *Proc. Roy. Soc. A.* **282**, 353 (1964).

[26] A. Franks, in *X-Ray Optics and X-Ray Micro-analysis*, eds. H. H. Pattee, V. E. Cosslett, and A. Engström (Academic, New York, 1963), p. 199.

[27] H. Volkmann, *Optik* **21**, 385 (1964).

[28] G. R. Harrison and G. W. Stroke, *J. Opt. Soc. Am.* **50**, 1153 (1960).

[29] G. R. Harrison, in *Molecular Structure and Spectroscopy*, Intern. Symposium Tokyo, 1962 (Butterworths, London, 1963).

[30] G. W. Stroke, *Phys. Letters* **5**, 45 (1963).

[31] R. F. Jarrell and G. W. Stroke, *Appl. Optics* **3**, 1251 (1964).

[32] J. A. R. Samson, *J. Opt. Soc. Am.* **52**, 525 (1962).

[33] E. M. Reeves and W. H. Parkinson, *J. Opt. Soc. Am.* **53**, 941 (1963).

[34] E. T. Arakawa, D. C. Hammer, and R. D. Birkhoff, *Appl. Optics* **3**, 79 (1964).

[35] W. F. Hanson and E. T. Arakawa, *J. Opt. Soc. Am.* **56**, 124 (1966).

[36] D. H. Tomboulian and W. E. Behring, *Appl. Optics* **3**, 501 (1964).

[37] G. Sprague, D. H. Tomboulian, and D. E. Bedo, *J. Opt. Soc. Am.* **45**, 756 (1955).

[38] A. H. Gabriel, J. R. Swain, and W. A. Waller, *J. Sci. Instr.* **42**, 94 (1965).

[39] K. Watanabe, *J. Opt. Soc. Am.* **43**, 318 (1953).

[40] G. B. Sabine, *Phys. Rev.* **55**, 1064 (1939).

[41] D. O. Landon, *Appl. Optics* **2**, 450 (1963).

[42] R. W. Hendrick, *J. Opt. Soc. Am.* **47**, 165 (1957).

[43] G. L. Johnson and R. F. Wuerker, in *X-Ray Optics and Microanalysis*, ed. H. H. Pattee, V. E. Cosslett, and A. Engström (Academic, New York, 1963), p. 229.

[44] A. P. Lukirskii, E. P. Savinov, and Y. F. Shepelev, *Optics and Spectroscopy* **15**, 290 (1963).

[45] A. P. Lukirskii, T. M. Zimkina, and I. A. Brytov, *Optics and Spectroscopy* **15**, 372 (1963).

[46] B. L. Henke in *X-Ray Microscopy and X-Ray Microanalysis*, eds. A. Engström, V. E. Cosslett, and H. H. Pattee (Elsevier, New York, 1960), p. 10.

[47] A. P. Lukirskii, E. P. Savinov, O. A. Ershov, and Y. F. Shepelev, *Optics and Spectroscopy* **16**, 168 (1964).

[48] R. P. Madden and L. R. Canfield, *J. Opt. Soc. Am.* **51**, 838 (1961).

[49] G. F. Jacobus, R. P. Madden, L. R. Canfield, *J. Opt. Soc. Am.* **53**, 1084 (1963).

[50] L. R. Canfield, G. Hass, and W. R. Hunter, *J. de Physique* **25**, 124 (1964).

[51] E. M. Reeves and W. H. Parkinson, *J. Opt. Soc. Am.* **53**, 941 (1963).

[52] G. Hass and R. Tousey, *J. Opt. Soc. Am.* **49**, 593 (1959).

[53] G. Hass, W. R. Hunter, and R. Tousey, *J. Opt. Soc. Am.* **46**, 1009 (1956); **47**, 1070 (1957).

[54] P. H. Berning, G. Hass, and R. P. Madden, *J. Opt. Soc. Am.* **50**, 586 (1960).

[55] L. R. Canfield, G. Hass, and J. E. Waylonis, *Appl. Optics* **5**, 45 (1966).

[56] W. R. Hunter, *Optica Acta* **9**, 255 (1962).

[57] S. Robin, *Compt. Rend. Acad. Sci.* **236**, 674 (1953).

[58] L. R. Canfield and G. Hass, *J. Opt. Soc. Am.* **55**, 61 (1965).

[59] O. P. Rustgi, W. C. Walker, and G. L. Weissler, *J. Opt. Soc. Am.* **51**, 1357 (1961).

[60] W. C. Walker, O. P. Rustgi, and G. L. Weissler, *J. Opt. Soc. Am.* **49**, 471 (1959).

[61] R. P. Madden, in *Physics of Thin Films*, ed. G. Hass (Academic, New York, 1963), p. 123.

3

Vacuum Spectrographs
and Monochromators

The dispersing element of a spectrograph is normally a prism or a diffraction grating. However, although spectrographs have been made using lithium fluoride prisms enabling the spectrum to be studied down to about 1100 Å, we will confine the discussion of this chapter to diffraction grating vacuum spectrographs and monochromators. The advantages of the grating spectrograph are wider wavelength coverage, higher resolving power, less scattered light, nearly linear dispersion (in the case of normal incidence), and precise wavelength measurements. Moreover, few if any vacuum prism spectrographs are available commercially, whereas numerous grating spectrographs and monochromators are manufactured.

3.1 SPECTROGRAPHS

The basic principle of the concave grating spectrograph lies in the focusing properties of the concave diffraction grating. As discussed in Chapter 2, these properties are such that the diffracted images of the source are sharply focused on a circle, called the Rowland circle, of a diameter equal to the radius of curvature of the grating, provided the surface of the grating is tangential to the Rowland circle, the ruled lines are at right angles to the Rowland plane, and the illuminated entrance slit is on the Rowland circle and parallel to the ruled lines of the grating. These conditions are essential in providing the highest possible resolving power. Figure 3.1 shows the basic optical layout of a spectrograph. The absolute values of the wavelengths can be determined from the following equations:

GRATING EQUATION, $\quad \pm m\lambda = d\,(\sin\alpha + \sin\beta);$ \hfill (3.1)

RECIPROCAL DISPERSION, $\quad \dfrac{d\lambda}{dl} = \dfrac{d}{mR}\cos\beta,$ \hfill (3.2)

43

where the positive sign in (3.1) applies to the "inside orders," and the negative sign to the "outside orders." The signs of α and β are opposite when they lie on different sides of the grating normal.

There are two basic types of spectrograph, the normal incidence spectrograph suitable for studies from 300 to 2000 Å and the grazing incidence spectrograph essential for studies below 300 Å. Spectrographs are

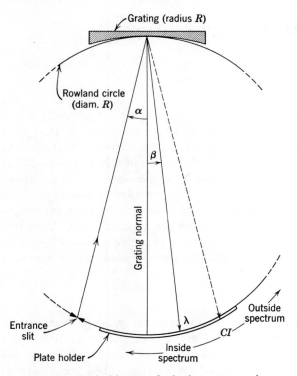

Fig. 3.1 Optical layout of a basic spectrograph.

generally referred to by the size of grating and the type of mount. Thus an instrument using a grating with a radius of curvature R of 1 m and with an angle of incidence close to zero is known as a 1 meter normal incidence spectrograph.

Normal Incidence Mount. When the angle of incidence α is less than approximately 10°, the radiation is considered to be directed at normal incidence to the grating. For α less than 10°, there is very little astigmatism and essentially no change in the reflectance, hence efficiency, of the grating. The actual angle of incidence, then, is dictated by the physical problem of mounting the photographic plate holder and light source.

The inside spectrum is normally used either in its first or higher orders since there is actually a loss in resolving power as β increases (see Chapter 2, Sect. 2.3). There is also a considerable saving in space when the inside spectrum is used.

Figure 3.2 shows a typical normal incidence spectrograph. The positions of the first order wavelengths for a 1200 line per mm grating are indicated when the angle of incidence $\alpha = 10°$. For these parameters, the astigmatism at 500 Å increases the image length of the slit by approximately $\frac{1}{30}$

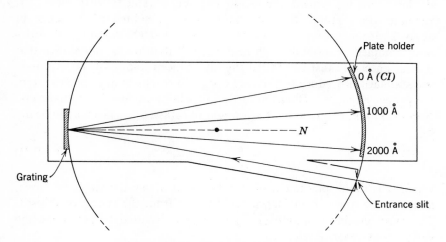

Fig. 3.2   Normal incidence spectrograph.

the height of the rulings. For 3 cm long rulings this amounts to only a 1 mm increase in the image length. Thus the astigmatism is very low for near normal incidence.

To obtain high resolution it is usual to work with higher spectral orders. With the simple normal incidence mount shown in Fig. 3.2, high orders cannot be obtained without adjusting the angle of incidence. This generally requires a focusing adjustment of the grating and the plateholder since it is desirable to keep the entrance slit fixed. If the entrance slit and the spectrum under study practically overlap, that is, $\alpha \sim \beta$, a much more compact instrument can be built. The focusing adjustments are also easier to perform. This is the principle of the Eagle mount.

Eagle Mount [1,2.]   In the Eagle mount, $\alpha \sim \beta$, and for this reason, the original Eagle design required the slit and plateholder to be located at some distance above and below the Rowland plane ("off-plane" mount) in order to avoid physical interference with each other. The angle of incidence is not restricted. However, when α is small, the normal incidence mount approximates to that of an "in-plane" Eagle mount with the spectrum viewed close to the entrance slit.

It would appear that the "off-plane" Eagle mount would adversely affect the resolving power of the spectrograph since the slit and plateholder are no longer on the Rowland circle. It has been shown by Wilkinson [3] and Namioka [4], however, that high resolution can still be obtained when the angular deviation of slit and plateholder is small.

Presumably, then, for a fixed separation between the slit and plateholder, the larger the instrument, the less defocussing obtained. With their 21 ft vacuum spectrograph, the slit and plateholder subtended an angle of about 2° at the grating. Resolving powers of 300,000 in the fourth order were obtained. To obtain high resolution, Wilkinson [3] has shown that the entrance slit must be rotated slightly. Since rotation of the slit turret is a standard proceduce to produce optimum results in any spectrograph, the proper angle of inclination of the slit will be obtained automatically. Namioka [4] has shown that the amount of rotation φ is given analytically by

$$\varphi = \frac{z_0}{R} \tan \alpha \sec \alpha, \tag{3.3}$$

where z_0 is the displacement of the slit center above the Rowland circle, and α is the angle of incidence for the rays from the center of the slit. The exact grating equation (2.15) must now be used to determine the various wavelengths; that is,

$$\frac{m\lambda}{d} = \left(1 + \frac{z_0^2}{R^2 \cos^2 \alpha}\right)^{-\frac{1}{2}} (\sin \alpha + \sin \beta). \tag{3.4}$$

The optimum width of a grating in the Eagle mount is given by [4]

$$W_{\text{opt}} = 2.12 \left[\left(\frac{4z_0^4}{\cos^4 \alpha} + \frac{2R^3 \lambda \cos \alpha \cos \beta}{\sin^2 \alpha \cos \beta + \sin^2 \beta \cos \alpha} \right)^{\frac{1}{4}} - \frac{2z_0^2}{\cos^2 \alpha} \right]^{\frac{1}{2}}. \tag{3.5}$$

Figure 3.3 shows a typical arrangement of optical parts in the off-plane Eagle mount. The light source is positioned at the entrance slit S, which is at the center of the plateholder. The vertical separation between S and the plateholder is $2z_0$. When the grating is at the position G, the central image falls at the center of the plateholder, allowing the instrument to be focused with ease. To change the wavelength interval under study, G is rotated about a vertical axis until the wavelength region of interest falls on the plateholder. Then G is moved along the track GS to the position G'. That is, the Rowland circle has been rotated about the fixed point S and the grating moved such that $\alpha = \beta$. The plateholder is then rotated

about *S* until it is back on the Rowland circle. For work at high orders
of the spectrum, the adjustments again
can be made in the visible. For example, the fourth order of the 1200 to
1500 Å region can be focused in air by
using the mercury 5461 Å line. Final
adjustments are made under vacuum by
using appropriate sharp emission lines.

The largest concave grating vacuum
spectrograph is the 35-foot (10.7 m) in-
plane Eagle mount. This instrument was
first designed and built by Douglas [5,6].
It is now built commercially by the
Jarrell-Ash Company (see Fig. 3.4). The
ruled width of the 1200 line per mm
grating is 7 in. This gives a reciprocal
dispersion of 0.8 Å per mm in the first

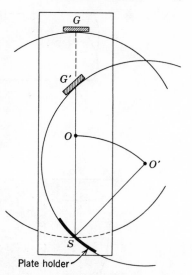

Fig. 3.3 Off-plane Eagle mount.

order while a resolution of 200,000 is reported at 1800 Å. The commercial instrument has a wavelength range of approximately 400 to 8000 Å.

The spectrograph consists of a vacuum tank 50 in. in diameter on one
end and tapering along its 40 foot length to 36 in. The vacuum tank,
unlike those in most large spectrographs, does not support the structural
optical assembly. Instead, the optics are mounted on concrete piers (see
Fig. 3.5) that, in turn, are independent of the building within which the
instrument is housed. Large stainless steel posts pass through the vacuum
wall by means of vacuum-tight flexible couplings. The optics are thus
independent of any distortion introduced into the vacuum vessel when the
unit is pumped as well as any vibration from pumps or external sources.

Fig. 3.4 10.7 m in-plane Eagle spectrograph (courtesy Jarrell-Ash Company).

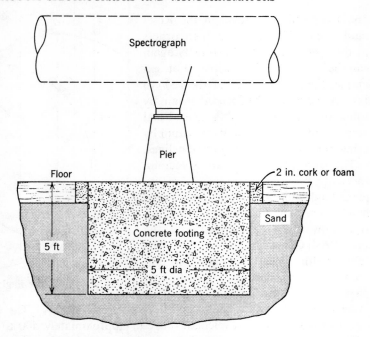

Fig. 3.5 Typical spectrograph supports to reduce vibration.

To obtain the highest resolution in such large spectrographs, the grating temperature must be held constant to within 0.05°C.

Grazing Incidence Mount.   The decrease in reflectance with decreasing wavelength of all grating materials necessitates the use of grazing incidence spectrographs for wavelengths below 200 to 300 Å. By making use of the total reflection experienced at extreme grazing incidence, Tyrén [8], using an angle of incidence of 89°, observed wavelengths as short as 12 Å. Kirkpatrick [8], using a grazing angle of 51′ of arc, photographed x-ray spectral lines down to 7 Å. The theoretical minimum wavelength for this grazing angle on glass is 5.6 Å (see (2.48)). As discussed in Chapter 2, the theoretical wavelength cutoff is lowered by using a reflecting surface of gold or platinum. Gabriel et al. [9] have recorded the shortest wavelength yet, namely 4.7 Å, by using a platinized grating at a grazing angle of 2°. For a platinum coating, the critical wavelength (in angstroms) is approximately equal to 2.64 times the grazing angle in degrees. Therefore, at a grazing angle of 2°, the wavelength cutoff should be approximately 5.3 Å. The optical arrangement for a grazing incidence mount is shown in Fig. 3.6.

To disperse shorter wavelengths the diffraction grating must be replaced by a crystal. The regular lattice spacing in the crystal now takes the place of the ruled lines of the grating. The grating equation, according to Bragg's law, is

$$m\lambda = 2d \sin \theta, \tag{3.6}$$

where θ is the grazing angle, m is the order number, and d is the distance between reflecting planes of the crystal. Thus the maximum wavelength diffracted by a crystal is $\lambda = 2d$. Synthetic crystals of potassium acid phthalate have been made with lattice spacings $d \sim 13$ Å and have proved useful for diffraction of wavelengths as long as 25 Å [10]. Recently, new crystals with larger interplaner spacings have been produced that are suitable for soft x-ray spectroscopy in the region 25 to 100 Å [11a].

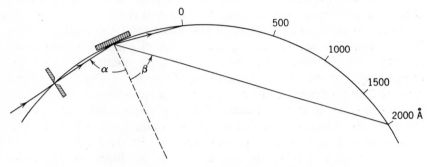

Fig. 3.6 Grazing incidence spectrograph.

These crystals are organic esters such as octadecyl hydrogen maleate ($2d \sim 64$ Å), dioctadecyl adipate ($2d \sim 94$ Å), and octadecyl hydrogen succinate ($2d \sim 97$ Å). The cleavage of these crystals is similar to mica. Although the crystals are relatively soft, they have sufficient mechanical strength to be handled without difficulty and can be easily bent to provide a degree of focusing. X-ray wavelengths are usually measured in x-ray units (xu) based on the lattice spacing of NaCl. At present, these x-ray units are 0.202 per cent smaller than the angstrom (10^{-8} cm).

Although the angular dispersion increases at grazing incidence, the resolving power actually decreases. This can be seen from the fact that the resolving power is proportional to W_{opt}, the optimum width of the grating, which, as shown in Table 2.2, rapidly decreases as the angle of incidence increases. For small gratings, say of 1 m radius, the optimum width and hence resolution is down by almost a factor of 10 at 88° compared with normal incidence. However, for larger instruments, since there is a practical limit to the width a grating can be ruled, the discrepancy in resolution becomes less.

Comparing the performance of a 6.8 m grazing incidence spectrograph ($\alpha = 82°$) with a 6.8 m Eagle mount, where both use 1200 line per mm gratings in the first order at 1000 Å, we see that the reciprocal dispersion in the grazing incidence mount is 0.64 Å per mm, while it is 1.28 Å per mm for the Eagle mount; a factor of 2 better at grazing incidence. The optimum width of the grating at 82° and 1000 Å is about 11 cm, whereas in the Eagle mount it is about 50 cm. However, the largest gratings available at present have ruled widths of only 20 cm, and this, then, limits the resolution obtained by the Eagle mount. Nevertheless, the resolution is a factor of 2 better than in the grazing incidence spectrograph and essentially compensates for the smaller dispersion. It is, of course, assumed that the total grating area is illuminated in both cases and that the slit widths are infinitely narrow. Thus it would appear that the performances of the two instruments are comparable *in the first order*.

The actual choice of instrument for work above 500 Å depends on many factors. For example, the grazing incidence spectrograph discussed above requires a volume twice that of the off-plane Eagle mount. Of course, the overall length of the grazing incidence spectrograph is about 12 ft compared to 21 ft for the Eagle mount. This may be an important factor where laboratory space is at a premium! With the given volume the grazing incidence spectrograph is limited in spectral range making it difficult or impossible to work in higher spectral orders. With the Eagle mount, on the other hand, it is relatively simple to study the fourth or fifth order spectrum. Finally, the astigmatism is very much greater at grazing incidence requiring much greater precision and care in the focusing of the spectrograph. Although there is an increase in the overall dispersion, it would appear that the major advantage of the grazing incidence mount is in enabling the spectrum to be studied from 400 Å down to the soft x-rays.

Echelle Mount. *Echelle* in French means "ladder" or "pair of steps." The diminutive form is *echelette*, while a rung of a ladder is *echelon*. These three words have all been used to describe regular steps in a reflecting material causing the diffraction of light.

The earliest form was Michelson's [12,13] *echelon*, which consisted of a series of optically flat glass plates of identical widths stacked together, but offset slightly to form a "ladder." Owing to the difficulties involved in preparing identical plates with the required precision, the echelon seldom has more than 40 plates and is extremely expensive to construct.

R. W. Wood [14] originated the shaped grooves of a diffraction grating (plane or concave) and called the grating an *echelette*. The grooves were shaped to increase the energy diffracted in a given order. This is the basis for the present day "blazed" gratings.

G. R. Harrison [15–17] designed an instrument with characteristics intermediate between the echelette and the echelon and called it an *echelle* grating. The echelle is essentially a ruled echelon. The precision of the grooves must approach that of the echelon. The echelle generally has more steps than the echelon, but fewer than the echelette.

Resolving powers of 1 million have not yet been achieved with a concave diffraction grating. At 1000 Å this represents a resolution of 0.001 Å. To obtain resolving powers as high or even higher, it is necessary to use the echelle mount. This consists of a plane diffraction grating with carefully shaped grooves used in the Eagle-type mount with $\alpha = \beta \approx 90°$, and with a concave mirror to focus the image. The same mirror may also be used to render the incident light parallel since the echelle grating must be used with parallel light. High resolution is obtained by working in high spectral orders, typically $m \sim 1000$. This means that some form of predisperser is necessary to separate overlapping orders.

Where only a narrow spectral range is under study, a predisperser can be used to isolate the region of interest. Then an echelle must be chosen with a free spectral range equal to the band pass given by the predisperser. The free spectral range F is given by

$$F = \frac{\lambda}{m},\tag{3.7}$$

where $m\lambda \sim$ twice the groove separation, as proved below (see 3.10). Thus for 1000 Å in its 1000th order, the free spectral range is only 1 Å. The band pass of the predisperser must therefore be ≤ 1 Å. The echelle in this case should have only 20 grooves per mm. The predisperser aids also in reducing scattered light.

On the other hand, when the echelle is crossed with a concave grating (i.e., the rulings of each grating are perpendicular to each other), separation of overlapping orders can be achieved with the additional advantage that a very large spectral range can be covered in a very small area. This is because the echelle produces the primary spectrum in one plane, whereas the concave grating separates the overlapping orders in a plane at right angles to that of the primary spectrum.

With the increased number of reflections necessary, the echelle mount has been used only above 1000 Å. Figure 3.7 shows a typical echelle spectrograph with a concave grating predisperser to separate higher orders.

Diffraction gratings normally achieve high resolution and dispersion by using a large number of grooves and working in moderately high orders, $m \sim 4$ to 20. From the grating equation $m\lambda = d(\sin \alpha + \sin \beta)$ and from the resolving power $\mathcal{R} = (m/d)W_{opt}$, we obtain

$$\mathcal{R} = \frac{W_{opt}}{\lambda}(\sin \alpha + \sin \beta).\tag{3.8}$$

For a given λ the resolving power increases when α and β are on the same side of the grating normal, whereas W_{opt} is independent of the sign of β. Further, $\mathcal{R}$ increases to a maximum as α and β increase, then decreases to zero at α or $\beta = 90°$. If we let $\alpha = \beta$ in (3.8) and differentiate $\mathcal{R}$ with respect to β, we find that $\mathcal{R}$ is a maximum at $\alpha = \beta = 54°45'$. The dispersion, on the other hand, is proportional to $1/\cos \beta$ and therefore continues to increase as β increases. Dispersion is the more important

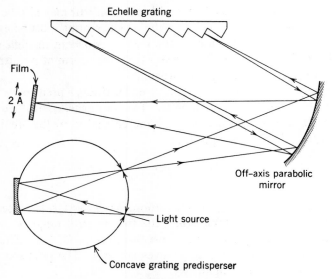

Fig. 3.7 Echelle spectrograph with predisperser.

quantity when resolution is limited by the practical slit width necessary to record a spectrum.

For a plane diffraction grating there is no optimum width. Thus from (3.8) the maximum resolving power is achieved when $\alpha = \beta \approx 90°$, namely,

$$\mathcal{R}_{max} = 2\,\frac{W}{\lambda}\,. \tag{3.9}$$

In this form we see that the maximum resolving power for a given wavelength is simply proportional to the width of the grating illuminated (and quite independent of the number of grooves).

For a grating 5 cm wide a resolving power of 10^6 should be obtained at 1000 Å. This suggests that it should be possible to rule a few properly shaped grooves so that most of the diffracted energy is returned into the desired order. From Chapter 2 we saw that the maximum energy went into the wavelength diffracted in the same direction as the specularly reflected beam from the surface of the facets, that is, for $2\theta = (\alpha + \beta)$. Under the conditions for maximum resolution this becomes $\theta = \alpha = \beta \approx 90°$. Figure 3.8 illustrates the echelle grating and shows its similarity to Michelson's reflecting echelon. In fact, the principle of the echelle is identical to that of the echelon. The difference lies only in the number of steps and in their production. With the parameters in Fig. 3.8, the grating equation $m\lambda = d(\sin\alpha + \sin\beta)$ of the echelle becomes

$$m\lambda = 2t - s\theta, \tag{3.10}$$

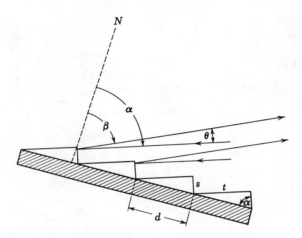

Fig. 3.8 Echelle grating. t is the width and s is the height of each step of the echelle.

which is the equation for the echelon, while the angular dispersion is given by

$$\frac{d\theta}{d\lambda} = \frac{m}{s}. \tag{3.11}$$

The ruling conditions for an echelle grating are very severe. The reflecting surfaces s must be optically flat and the separation t between steps must be held to within $\frac{1}{100}\lambda$. The inaccuracies in spacing of the grooves greatly decrease the actual resolving power of a grating at high angles of incidence. The ruling of the gratings are controlled by interferometric methods to achieve such perfection.

Further dispersion can be achieved by using two echelles in tandem. Although this has been successful in the visible [17,18], every extra reflection in the vacuum ultraviolet region produces a large decrease in intensity.

The echelle grating and the echelon appear to be the only instruments capable of producing resolving powers in excess of 10^6 in the vacuum uv.

Separation of the overlapping orders can be achieved by crossing the echelle and echelon with a concave diffraction grating.

Ebert-Fastie Mount.   Another plane grating mount (similar to the echelle mount) with three reflections was designed by Ebert [19] in 1889 and rediscovered and modified by Fastie [20,21] in 1952. Although it has been used mainly for the near ultraviolet and visible region of the spectrum, it should be useful down to 1100 Å when the reflecting components are coated with Al-MgF$_2$ films.

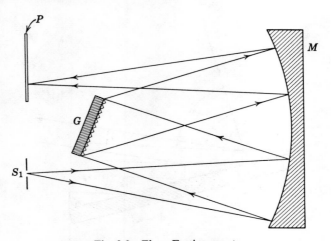

Fig. 3.9   Ebert-Fastie mount.

The optical system is shown in Fig. 3.9. Light from the entrance slit S_1 is rendered parallel by the spherical concave mirror M and is reflected onto the plane grating G. The diffracted rays are then refocused by M onto the photographic plate P.

In its original form, as described by Ebert, the center of the grating lay in line with S_1 and P. In this position, however, only a small spectral range was in focus at the plate, this portion being situated at the same distance as S_1 from G, but on the opposite side. The focal plane is actually curved. If G is located at a distance from M of about 0.8 times that of S_1, the focal plane of the spectrum will lie in a straight line.

Instead of the single mirror, two smaller ones could be used as in the Czerny-Turner system [22]. Fastie has shown that a very high resolution can be achieved with this mount.

Miscellaneous Mounts.   All the standard optical mountings of concave gratings, such as the Rowland, Paschen-Runge, and Abney mounts, used in the visible and infrared region are all applicable to the vacuum ultraviolet. However, they simply describe various mechanisms for changing the spectral region to be photographed; the spectrographs themselves are still based on the Rowland circle.

A somewhat different optical arrangement of the concave grating is its use with parallel light. This type of mounting was first described in 1896 by Wadsworth [23], who found that stigmatic images could be obtained from a concave grating when viewed near the normal to the grating. From

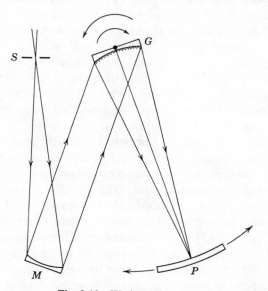

Fig. 3.10   Wadsworth mount.

Sect. 2.1, it is shown that the images are formed at a distance r' from the grating given by

$$r' = \frac{R}{1 + \cos \alpha}. \tag{3.12}$$

This is the equation of a parabola. For a given α, a stigmatic spectrum is generated in the vicinity of the normal, where the photographic plate has an approximate curvature of a circle with radius $R/(1 + \cos \alpha)$. In changing the spectral range, α is varied; thus the curvature of the plate holder must be changed. Though truly stigmatic images are obtained only on the normal, for most purposes a range of several hundred angstroms can be obtained at one setting.

Figure 3.10 shows the arrangement of the Wadsworth mount as used by Meggers and Burns [24]. The spectral range is varied by rotating the grating G about its center. The plateholder is attached to and free to move along the rod GP, which is also the normal to G.

The Wadsworth mount has been used in the construction of a simple spectroheliograph for photographing a stigmatic image of the sun at the He II 304 Å line [25].

3.2 MONOCHROMATORS

As in the vacuum spectrograph, the use of a concave grating as the dispersing device requires that the optical components of the monochromator always lie on the Rowland circle to produce perfect focus. The optical parts in such a case are the entrance and exit slits and the concave grating. To maintain perfect focus and still achieve monochromatic radiation from the exit slit, either one element must move along the Rowland circle, or else the Rowland circle must rotate about one of the elements. The other two elements are moved appropriately so that they stay on the circle. Several mountings described below achieve monochromatic action and still maintain focus.

There are many situations in which a monochromator with perfect focus is not so important as such overriding considerations as fixed exit or entrance slits, size, cost, an undeviating exit beam, and others. Thus there is probably no ideal monochromator. The selection of a design is dictated generally by the application. Certainly the simplest and most straight forward monochromator would be one in which the grating was simply rotated about its center. If the instrument initially had the optical elements on the Rowland circle for the wavelength in the middle of the spectral range under study, the defocusing would be minimized.

Some of the more important monochromator designs are discussed below. All of these designs attempt to counteract any defocusing of the instrument.

Rowland Circle Mounting

Grazing Incidence. In the grazing incidence mounting any deviation of the slits or the grating from the Rowland circle results in a very large degree of defocusing. Thus monochromatic action at grazing incidence is confined to moving one or more of the optical elements along the Rowland circle.

One of the earliest vacuum monochromators and the first grazing incidence monochromator was constructed by Baker in 1938 [26]. In his instrument both slits were fixed and the grating was constrained to move on the Rowland circle. In this geometry both the angles of incidence and diffraction change, but the angle subtended by the two slits at the center of the grating remains constant; that is, $(\alpha - \beta) =$ constant (remembering that the sign of β is opposite to that of α if it lies on the opposite side of the grating normal). The angular dispersion in this case is very nearly constant over a limited range of β and is given by

$$\frac{d\lambda}{d\beta} = 2d \cos \varphi \cos (\varphi + \beta), \qquad (3.13)$$

where φ is a constant equal to $\frac{1}{2}(\alpha - \beta)$. If fixed exit and entrance slits are required, then this is the only appropriate mounting.

Piore et al. [27] used a movable exit slit and were thereby able to make use of the increasing dispersion at shorter wavelengths. Landon [28] describes a similar instrument, but his scanning mechanism allows a linear

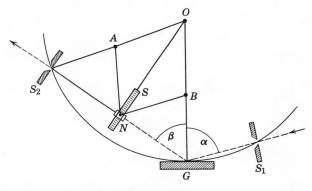

Fig. 3.11 Scanning mechanism to produce a linear wavelength readout. The wavelength is proportional to the length ON.

wavelength readout. Figure 3.11 illustrates his scanning system. A rigid rod OG is attached to the grating G, while a movable rod OS_2 is pivoted at O, the center of the Rowland circle, and fixed to the exit slit S_2. A precision screw S is aligned to bisect the isosceles triangle OGS_2, and a rigid rod NS_2 is fixed to the precision nut N and to the exit slit. Two supporting rods are also fixed to N and pivoted at A and B, the midpoints of OS_2 and OG, respectively. Rotation of the screw constrains the exit slit to move along the Rowland circle. The length of the screw ON is always equal to $OG \sin \beta$. However, since the angle of incidence is constant, the wavelength is also proportional to $\sin \beta$. Thus the length of ON is proportional to the wavelength, and a linear wavelength scale is obtained that can be read out on a counter.

A novel method of moving the exit slit has been used by Hinteregger [29,30] and associates in a rocket monochromator for scanning the solar spectrum from about 1300 to 60 Å. Their method consisted of cutting several exit slits in an endless steel band that was constrained to move along a segment of the Rowland circle. The radiation was detected by a fixed electron multiplier with a cathode sufficiently large to receive the sweeping exit beam. A schematic of the monochromator is shown in Fig. 3.12.

The principal use for this type of monochromator is, of course, in the study of spectra from various sources, not in its ability to provide mono-chromatic radiation for other research projects.

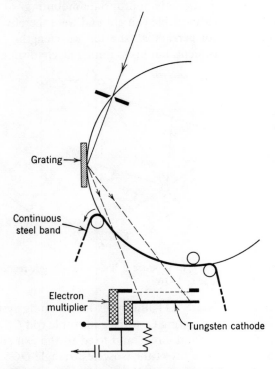

Fig. 3.12   Hinteregger-type scanning monochromator.

Another instrument utilizing the moving exit slit has been produced by the McPherson Instrument Corp. [31]. In this case the exit slit is mounted on a turntable that moves along a precision machined track coinciding with the Rowland circle. The exit and entrance slits are readily inter-changeable. For the movable slit to be accessible outside the vacuum housing, the slit is connected to the vacuum system through a series of telescoping tubes and O-ring seals. Figure 3.13 is a photograph of a 2 m McPherson grazing incidence monochromator shown with the exit slit mounted on the track and the telescoping tubes.

A grazing incidence monochromator in which the Rowland circle is pivoted about the exit slit keeping the angle of incidence constant has been constructed by Salle and Vodar [32]. Figure 3.14 shows a typical arrangement of their scanning mechanism. The grating G is rigidly fixed to the entrance slit S_1 at a fixed angle of incidence α. The length of the rod GS_1 is, therefore, $R \cos \alpha$. Both G and S_1 will subtend a constant angle of $(90 - \alpha)$ at S_2. The grating is mounted on a turntable constrained to move

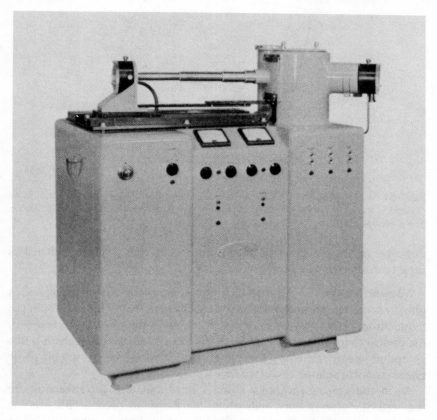

Fig. 3.13 McPherson 2 m grazing incidence scanning monochromator (courtesy McPherson Instrument Corporation).

along the track GS_2, and S_1 is constrained to move along the track S_1S_2. Thus any movement of S_1 along S_1S_2 will result in a larger displacement of the grating along S_2G coupled simultaneously with a rotation of the grating relative to S_2G. Because GS_1 and α are constants, it is easily seen that any movement of S_1 along S_1S_2 still maintains S_1 and G on the Rowland circle.

This mounting is particularly advantageous in providing a constant angle of incidence and, at the same time, a fixed exit slit and a fixed

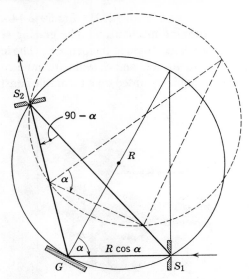

Fig. 3.14 Vodar grazing incidence monochromator. Scanning mechanism maintains the optical elements on the Rowland circle and maintains a constant angle of incidence. The Rowland circle is pivoted about the exit slit S_2.

direction of the emergent beam. Furthermore, the dispersion increases rapidly to shorter wavelengths.

Normal Incidence. All the techniques described above to create a grazing incidence monochromator can also be applied to normal incidence monochromators. To obtain precise focusing as the spectrum is scanned, the optical elements must lie on the Rowland circle. Actually, there is one exception to this rule when $\alpha = \beta$. This is the condition for the off-plane Eagle monochromator discussed below.

As in the grazing incidence case, if fixed exit and entrance slits are required, the grating must slide along the Rowland circle. This was the technique adopted by Tousey et al. [33] and by Fujioka and Ito [34] in their *radius mounting* [35]. The grating was mounted at the end of an arm pivoted about the center of the Rowland circle. The arm coincided with the normal to the grating. The motion of the grating causes the exit beam to change direction unless the exit beam is sufficiently baffled. This, of course, decreases the effective width of the grating available for dispersion and decreases the intensity of the exit beam.

To avoid the changing aspect of the grating as seen from the axes of both the light source and the detector Clarke and Garton [36] moved both the exit and entrance slits on tracks directed from the respective slits to the center of the grating. The grating was allowed to rotate about its center.

The grating and the two slits were linked by arms pivoted at the center of the Rowland circle O. Any displacement of O caused the grating to rotate and the slits to slide along their tracks, remaining automatically on the Rowland circle since the arms were radii of the Rowland circle.

McPherson [31] used a fixed entrance slit and allowed the grating and exit slit (or plateholder) to move in his 2-m combination vacuum spectrograph and scanning monochromator. Figure 3.15 shows the scanning

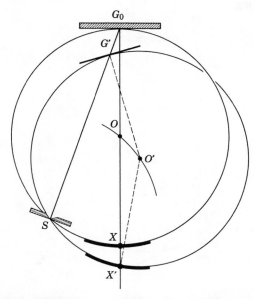

Fig. 3.15 Scanning mechanism for McPherson 2 m combination normal incidence spectrograph and scanning monochromator.

mechanism. In the initial position, the grating is at G_0 with the light from the fixed entrance slit S at an angle of incidence of $8°15'$, while the exit slit X is at the opposite end of a diameter passing through G_0. The plateholder is tangential to the diameter at X. Both the grating and X are rigidly attached to arms pivoted at O. Then by displacing O along the arc of a circle, with center S, G_0 and X slide along tracks G_0S and G_0X, respectively. In this arrangement there is a very slight change in the direction of the exit beam.

Off-Rowland Circle Mounting

The following monochromators have fixed exit and entrance slits, the scanning action being achieved by a simple rotation or by a simultaneous rotation and translation of the grating.

Normal Incidence. The class of monochromator described in this section appears to have been developed independently by a number of groups [37–41]. The basic principle is that the grating is constrained to move along the bisector of the angle subtended by the slits at the center of the grating; simultaneously the grating is rotated about a vertical axis tangent to its center. The rotation of the grating provides the monochromatic action, whereas the linear motion of the grating determines the degree of focusing.

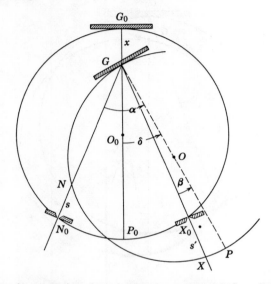

Fig. 3.16 Monochromator based on the rotation and translation of the grating along the bisector of the angle subtended by the slits at the grating center.

The linear displacement of the grating necessary to give the best focusing can be found as follows.

From the fundamental focal condition (2.9), we have

$$\frac{\cos^2 \alpha}{r} - \frac{\cos \alpha}{R} + \frac{\cos^2 \beta}{r'} - \frac{\cos \beta}{R} = 0,$$

where r and r' refer to the entrance and exit slit distances from the center of the grating. If now the entrance slit distance is increased a distance s from the Rowland circle along the line joining slit and grating, and if the exit slit distance is similarly decreased by an amount s', then (2.9) becomes (see Fig. 3.16)

$$\frac{\cos^2 \alpha}{s + R \cos \alpha} - \frac{\cos \alpha}{R} + \frac{\cos^2 \beta}{R \cos \beta - s'} - \frac{\cos \beta}{R} = 0. \qquad (3.14)$$

Solving for s', (3.14) becomes

$$s' = s\left[1 + \frac{s}{R}\left(\frac{\cos \alpha + \cos \beta}{\cos \alpha \cos \beta}\right)\right]^{-1}. \qquad (3.15)$$

When

$$\frac{s}{R}\left(\frac{\cos \alpha + \cos \beta}{\cos \alpha \cos \beta}\right) \ll 1, \qquad \text{then} \qquad s' \approx s,$$

which is approximately the case for near normal incidence. Referring to Fig. 3.16, we see that if the grating G_0, entrance slit N_0, and exit slit X_0 are initially on the Rowland circle such that the central image falls on X_0 then for a rotation of the grating through an angle δ and a linear motion through a distance x, the optical elements should now be at G, N, and X. Thus we see that the entrance slit lies a distance s outside the new Rowland circle, whereas the exit slit lies a distance s' inside the circle. The distance s is found as follows.

By construction, we have $GN_0 = GX_0$; that is

$$R \cos \alpha + s = R \cos \beta - s'. \qquad (3.16)$$

Let α_0 be the angle of incidence in the initial position. Then, if the change in the angle subtended by the slits at the grating is small over the spectral range scanned, (3.16) becomes, for $s = s'$,

$$s = R \sin \alpha_0 \sin \delta. \qquad (3.17)$$

To find the linear motion x of the grating it is noted that in Fig. 3.16, to a good approximation,

$$s = R \cos (\alpha - \delta) - R \cos \alpha - x \cos (\alpha - \delta) \qquad (3.18)$$

and

$$s' = -R \cos (\alpha - \delta) + R \cos \beta + x \cos (\alpha - \delta), \qquad (3.19)$$

where $R = G_0 P_0 = GP$ and $R - x = GP_0$. Now if $s = s'$, then by equating (3.18) and (3.19) and noting that $\delta = \frac{1}{2}(\alpha + \beta)$, we obtain

$$x = R(1 - \cos \delta). \qquad (3.20)$$

A more rigorous analysis using similar triangles $G_0 G N_0$ and $G_0 G X_0$ gives

$$x = \frac{R}{2} \sin^2 \delta - \frac{R}{2} \tan^2 \alpha_0. \qquad (3.21)$$

However, for δ small, (3.20) is sufficiently accurate. Moreover, it is the precise equation obtained if the Rowland circle is pivoted about P_0, the point of intersection of the grating normal with the Rowland circle in its initial position. The use of (3.20) as the focusing condition leads to a simple scanning mechanism. For example, if the grating is fixed to an arm lying along its normal and pivoted at the center of the Rowland circle O (Fig. 3.16), and if the arm OP_0 is also pivoted at O and at the fixed point P_0, then a simple movement of O will pivot the Rowland circle about P_0. The grating, of course, is constrained to move along the track G_0P_0. This type of mounting and scanning mechanism has been described and analyzed by Vodar [37] and Robin [38] and is produced commercially by the Jarrell-Ash Corporation*. A more rigorous analysis has been given by Pouey and Romand [42].

An empirical approach has been used by McPherson [39]. In his instrument† a lever arm of suitable length greater than $R/2$, but less than R, is rigidly attached to the grating mount. The arm coincides with the grating normal in the initial position. The end of the lever arm is constrained to follow a properly shaped cam. The curvature of the cam is determined empirically by focusing various wavelengths at the exit slit in the range 0–6000 Å and noting the position of the cam follower. These positions are found to lie on a smooth curve. The cam is then machined to this curvature. One advantage of this procedure is that the curvature of the cam automatically compensates for any errors in the original positioning of the grating. With slit widths of 10 μ and a 1 m, 600 line per mm grating, a resolution of approximately 0.23 Å has been obtained over the range 0–6000 Å.

Eagle off-plane Mounting.   This can be considered as a special case of the mounting described in the preceding section, for, when the entrance and exit slits coincide the bisector becomes simply the line joining the slits to the grating. In the off-plane Eagle mounting the entrance and exit slits are placed symmetrically above and below the Rowland plane, so that the distance z_0 between the center of the slit and the Rowland plane is the same for both slits. The angle of incidence α_0 for the rays from the center of the entrance slit is equal to the angle of diffraction β_0 for the rays that originate at the center of the entrance slit and are diffracted from the center of the grating rulings. Both α_0 and β_0 are measured on the Rowland plane.

To scan the spectrum the grating is rotated and moved in the direction of the fixed slits, whereas the Rowland circle pivots about the slits and no defocusing of the normal type occurs; that is, no element is forced off the circle as the spectrum is scanned. The question, however, is to what extent the monochromator already is defocused by locating the slits above and below the Rowland plane. As was discussed earlier in Sect. 3.1, Wilkinson [3] and Namioka [4] have shown that it does not defocus a

---

* Jarrell-Ash Corporation, Waltham, Mass.
† McPherson Instrument Company, Acton, Mass.

spectrograph appreciably when a large radius grating is used, provided the entrance slit is slightly rotated with respect to the rulings of the grating. Namioka has shown the following conditions to hold [43].

SLIT ROTATION
$$\varphi = \frac{z_0}{R} \tan \alpha \sec \alpha;$$
(3.22)

WAVELENGTH
$$\lambda = \frac{2d}{m}\left(1 + \frac{z_0{}^2}{R^2 \cos^2 \alpha}\right)^{-\frac{1}{2}};$$
(3.23)

OPTIMUM WIDTH OF GRATING

$$W_{\text{opt}} = 2.12\left[\left(\frac{4z_0{}^4}{\cos^4 \alpha} + \frac{\lambda R^3 \cot \alpha}{\sin \alpha}\right)^{\frac{1}{2}} - \frac{2z_0{}^2}{\cos^2 \alpha}\right]^{\frac{1}{2}}.$$
(3.24)

The symbols have been defined in Section 3.1. Typical values of λ, φ, and W_{opt} are approximately 591 Å, 2.5′ of arc, and 37 cm, respectively, for a 3 m grating with 1200 lines per mm used in the first order at an angle of incidence equal to 2° and with $z_0 = 6$ cm. It can be seen that the optimum width of the grating is much larger than that currently available on the market.

The scanning mechanism shown in Fig. 3.3 for the off-plane Eagle spectrograph is similar to that of McPherson's shown in Fig. 3.15. For the slits to be fixed and the grating constrained to move along tracks joining the grating and the slits, the center of the Rowland circle must move along an arc of a circle with center S and radius equal to that of the Rowland circle. The center O need not physically be moved along the arc, and any link mechanism which effects a virtual displacement of O along the arc will suffice and, indeed, would probably be more compact. Such a mechanism is described by Bair et al. [44] and by Namioka [43]. These scanning mechanisms could be used in the monochromator described above.

Johnson-Onaka Mounting. This mounting is very nearly identical to the normal incidence mountings described above. Scanning the spectrum past the exit slit is accomplished by rotating the grating about an off-axis pivot point. The point is chosen such that the grating is moved approximately along the bisector of the angle subtended by the entrance and exit slits at the grating. In the limit of small rotations the grating essentially moves along the bisector; hence the mounting is identical to those described above except that the mechanism for moving the grating is different. For larger rotations, of course, the grating center deviates more from the bisector.

Johnson [45] first described the use of an off-axis pivot point for the rotation of the grating when used near normal incidence. Onaka [46] analysed the mounting in detail and showed that it was applicable for any angle of incidence with the reservation that the spectral range of good focus decreases when the angle subtended by the slits at the grating ($= 2\varphi$) deviates from the most optimum values. These values of φ are approximately zero and 35°. It should be emphasized, however, that good focusing is achieved for any φ over several hundred angstroms.

To locate the pivot point Onaka proceeded as follows. In Fig. 3.17 the Rowland circle focusing conditions are shown when the average

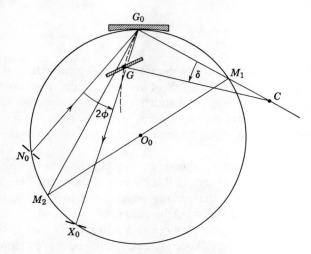

Fig. 3.17   Johnson-Onaka monochromator.

wavelength λ_0 is emerging from the exit slit X_0. The entrance slit is at N_0, and the grating at G_0. M_1 and M_2 are midway points on the Rowland circle between N_0 and X_0. The pivot point C is chosen to lie at the appropriate position on the line joining G_0 and M_1. The reason for the selection of this line is that it makes a right angle with the bisector G_0M_2, and for small rotations, the grating is moved towards M_2 approximately along G_0M_2. This movement reduces the amount of changes in the incident and emerging directions of the radiation. Thus the solution for the location of C is not a general one.

With the above constraint on the position of C, its location at a distance l from G_0 is found as follows. Using Beutler's equation for focusing and Rayleigh's criterion that the total pathlength for the various rays must not vary by more than $\lambda/4$, we get

$$\left| \frac{A}{2} W^2 + \frac{B}{2} W^3 + \text{higher order approximations} \right| \leq \frac{\lambda}{4}, \qquad (3.25)$$

where

$$A = \frac{\cos^2 \alpha}{r} - \frac{\cos \alpha}{R} + \frac{\cos^2 \beta}{r'} - \frac{\cos \beta}{R},$$

and

$$B = \frac{\sin \alpha}{r}\left(\frac{\cos^2 \alpha}{r} - \frac{\cos \alpha}{R}\right) + \frac{\sin \beta}{R}\left(\frac{\cos^2 \beta}{r'} - \frac{\cos \beta}{R}\right).$$

W is the half width of the grating, while the other terms are as previously defined. By expanding α, β, $1/r$, and $1/r'$ in a power series in δ about the initial position α_0 and β_0, we get for A and B,

$$A = A_1\delta + A_2\delta^2 + \cdots$$

and

$$B = B_1\delta + \cdots$$

where

$$A_1 = \frac{1}{R^2}(2l\cos\varphi - R\sin\alpha_0 - R\sin\beta_0 - l\tan\alpha_0\sin\varphi + l\tan\beta_0\sin\varphi),$$

$$(3.26)$$

and for small l, as is the case in the extreme ultraviolet,

$$A_2 = \frac{2 - 3\cos^2\varphi}{R\cos\varphi}, \tag{3.27}$$

and

$$B_1 = \frac{-2\sin^2\varphi}{R^2\cos\varphi}. \tag{3.28}$$

The ideal arrangement would be realized when all the coefficients $A_1, A_2, \ldots, B_1, \ldots$ were identically equal to zero. Unfortunately, this is not possible. Thus we set the largest coefficient, namely A_1, equal to zero. This gives the location of C as

$$l = \frac{R\sin\theta}{1 - \frac{1}{2}\tan\varphi\,(\tan\alpha_0 - \tan\beta_0)}, \tag{3.29}$$

where $\theta = \frac{1}{2}(\alpha_0 + \beta_0)$ and $\varphi = \frac{1}{2}(\alpha_0 - \beta_0)$.

The wavelength range for focus within the Rayleigh criterion is given by (3.25), which now becomes

$$\frac{W^2}{2}|A_2\delta^2 \pm B_1 W\delta| \le \frac{\lambda}{4}. \tag{3.30}$$

The $\pm$ sign is used since the center of the grating was taken as the origin in the derivation of (3.25). Therefore W, the half width of the grating, can be either positive or negative. For a simple rotation of the grating about its center,

$$\lambda = 2d \cos \varphi \sin (\theta + \delta) \tag{3.31}$$

and for small δ,

$$\lambda \simeq 2d \cos \varphi (\sin \theta + \delta \cos \theta). \tag{3.32}$$

Equation 3.32 is valid for the off-axis rotation when δ is small. Inserting (3.27), (3.28), and (3.32) into (3.30), a quadratic equation in δ is obtained,

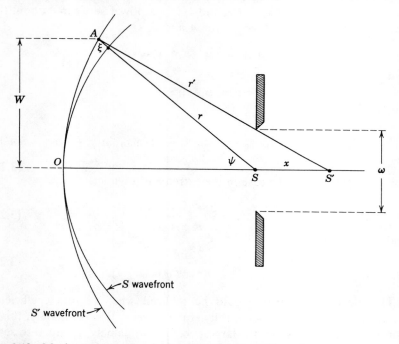

Fig. 3.18 Maximum separation ξ of wavefronts to provide optimum resolution for a slit of width ω.

the solution of which yields the permissible limits of rotation to maintain focus within the Rayleigh criterion.

In practice the Rayleigh condition is too severe and the limiting resolution is controlled by the necessary slit widths. Thus $\lambda/4$ is replaced by $\omega W/4r$, where ω is the slit width and $r \sim R \cos \alpha_0$ is the distance of the exit slit from the grating. The proof of this substitution is given as follows.

In Fig. 3.18 it is assumed that the correct focus is given when the light beam converges at S, and the maximum allowable defocusing occurs when the beam converges at S', where the width of the beam at S is just equal to the slit width. The separation between the two wavefronts at A

is equal to ξ and this is the quantity which replaces $\lambda/4$. Let the distance $AS = (r + \xi)$, $AS' = r'$, $SS' = x$, and the angle $A\hat{S}O = \psi$. Then

$$(r')^2 = (r + \xi)^2 + x^2 - 2x(r + \xi) \cos (180 - \psi), \qquad (3.33)$$

where $r' = r + x$, as can be seen when $\psi = 0$. Eliminating r' and ignoring terms in $x\xi$ and ξ^2, (3.33) becomes

$$\xi \approx x(1 - \cos \psi). \qquad (3.34)$$

For ψ small, $\cos \psi \sim 1 - \psi^2/2$, $\psi \sim W/r$, and $x \sim \omega/2\psi$. Therefore,

$$\xi \approx \frac{\omega W}{4r}. \qquad (3.35)$$

A monochromator with a 1 m 600 line per mm grating focused for 1000 Å and with slits subtending an angle of 9° with respect to the grating should give a useful spectral range from 370 to 1630 Å for a slit width of 40 μ and a grating width of 4 cm.

Seya-Namioka Mounting. By far the simplest scanning mechanism of any monochromator is the rotation of the grating about a vertical axis through the center of the grating. This method has been used in several cases for near normal incidence and for a limited wavelength scan [47,48]. It has proved to be satisfactory, provided a resolution of a few angstroms was acceptable. In 1952 Seya [49] analyzed the focusing conditions for a simple rotation of the grating and found that rather good focus could be expected over a large spectral range if the slits subtended an angle of approximately 70° at the grating. Namioka [50,51] constructed a monochromator based on Seya's calculations and later analyzed the mounting in more detail with emphasis on its astigmatism and resolving power. Figure 3.19 illustrates a commercial Seya-Namioka monochromator.

The mounting is analyzed as follows. Starting with (2.9), the focal equation for the concave grating,

$$\frac{\cos^2 \alpha}{r} - \frac{\cos \alpha}{R} + \frac{\cos^2 \beta}{r'} - \frac{\cos \beta}{R} = 0,$$

the constraint $\alpha - \beta = C$ is applied, where C is a constant. This represents the constant angle subtended by the slits at the center of the grating. α and β are of opposite sign when they lie on opposite sides of the grating normal. If we denote (2.9) by f and substitute $\alpha - C$ for β, then

$$f = \frac{\cos^2 \alpha}{r} + \frac{\cos^2 (\alpha - C)}{r'} - \frac{\cos \alpha}{R} - \frac{\cos (\alpha - C)}{R} = 0. \qquad (3.36)$$

It is more convenient to express f in terms of $\rho = R/r$ and $\rho' = R/r'$. Thus

$$f = \rho \cos^2 \alpha + \rho' \cos^2 (\alpha - C) - [\cos \alpha + \cos (\alpha - C)]. \qquad (3.37)$$

In this mounting ρ and C are constant. To maintain good focus as the grating is rotated, ρ' should also stay constant or at least change very slowly. The problem is now to select the best value of C and ρ that will make f as close to zero as possible and for which the change in ρ' is a minimum. The conditions for f to be approximately zero under the above constraints can be determined as follows.

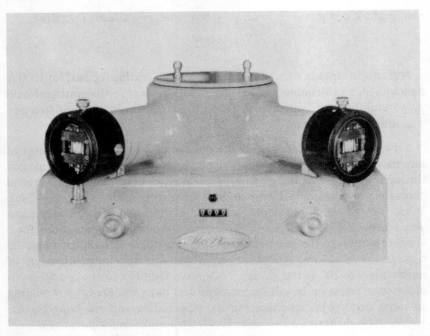

Fig. 3.19 $\frac{1}{2}$ m Seya-Namioka monochromator (courtesy McPherson Instrument Corporation).

If α_0 is the angle of incidence in the initial position of the grating, then for a rotation $\Delta\alpha$, the function f can be expanded by a Taylor's series around $\alpha = \alpha_0$. Thus,

$$f = f(\alpha_0) + f'(\alpha_0)\,\Delta\alpha + f''(\alpha_0)\,\Delta\alpha^2 + \cdots, \qquad (3.38)$$

where f' and f'' denote respectively the first and second derivatives of f with respect to α. When $\Delta\alpha \ll 1$, f is approximately zero when the coefficients $f(\alpha_0)$, $f'(\alpha_0)$, and $f''(\alpha_0)$ are identically zero; that is,

$$\rho \cos^2 \alpha_0 + \rho' \cos^2 (\alpha_0 - C) - [\cos \alpha_0 + \cos (\alpha_0 - C)] = 0; \quad (3.39)$$

$$-\rho \sin 2\alpha_0 - \rho' \sin 2(\alpha_0 - C) + [\sin \alpha_0 + \sin (\alpha_0 - C)] = 0; \quad (3.40)$$

$$-2\rho \cos 2\alpha_0 - 2\rho' \cos 2(\alpha_0 - C) + [\cos \alpha_0 + \cos (\alpha_0 - C)] = 0. \quad (3.41)$$

To satisfy these three simultaneous equations, the following determinant must be zero:

$$\begin{vmatrix} \cos^2 \alpha_0 & \cos^2 (\alpha_0 - C) & -[\cos \alpha_0 + \cos (\alpha_0 - C)] \\ -\sin 2\alpha_0 & -\sin 2(\alpha_0 - C) & [\sin \alpha_0 + \sin (\alpha_0 - C)] \\ -2 \cos 2\alpha_0 & -2 \cos 2(\alpha_0 - C) & [\cos \alpha_0 + \cos (\alpha_0 - C)] \end{vmatrix} = 0.$$

$$(3.42)$$

The solutions for (3.42) are

$$C_1 = \alpha_0 + \tan^{-1}(\sqrt{3} \sec \alpha_0 - 2 \tan \alpha_0); \tag{3.43}$$

$$C_2 = \alpha_0 - \tan^{-1}(\sqrt{3} \sec \alpha_0 + 2 \tan \alpha_0); \tag{3.44}$$

$$C_3 = 0. \tag{3.45}$$

For any given value of α_0, (3.43) and (3.45) give the values of C which satisfy the focal condition (3.36) within the approximation used in (3.38). The solution C_2 is omitted, however, since C_2 does not provide either vacuum ultraviolet or visible spectra.

To select the best value of C and α_0 it is necessary that the variation in C and ρ' with respect to α_0 be zero or at least a minimum. Consider C_3 first. Obviously $dC_3/d\alpha_0$ is zero, and the optimum value of C_3 therefore remains constant during the rotation of the grating. Inserting the value $C_3 = 0$ into (3.39) and taking the partial derivative of ρ' with respect to α_0 and equating to zero, we get

$$\left. \frac{\partial \rho'}{\partial \alpha_0} \right|_{r, C=\text{const}} = 2 \sec \alpha_0 \tan \alpha_0 = 0. \tag{3.46}$$

This is possible only for $\alpha_0 = 0$. Thus the solution C_3 provides $C = 0$. Therefore $\alpha = \beta$, and $\alpha_0 = 0$. By inspection we see that as the Rowland circle is rotated about the grating center, the exit and entrance slits both lie inside or both outside of the circle depending on the direction of rotation. This is not an optimum focusing condition as can be seen from the discussion on p. 62 above. Furthermore the change in ρ' is excessive compared with that obtained using the solution C_1, as is discussed below. Thus the solution C_3 also will be omitted.

Considering C_1, the condition for $dC_1/d\alpha_0 = 0$ is that $\alpha_0 = 35°15'$. Inserting this value into (3.43), the value of C is found to be $C = 70°32'$. Thus C is essentially equal to $2\alpha_0$.

From (3.39) and (3.40), we find

$$\rho' = \tfrac{1}{4} \sin 2\alpha_0 \sec (\alpha_0 - C) \csc C + \tfrac{1}{2} \sin \alpha_0 \csc C + \tfrac{1}{2} \sec (\alpha_0 - C). \tag{3.47}$$

Substituting $C = 2\alpha_0$ into (3.47), $\rho' = 1/\cos \alpha_0$. Similarly, $\rho = 1/\cos \alpha_0$. That is, using the optimum value of C, the initial position is such that the optical elements all lie on the Rowland circle and $\alpha_0 = -\beta_0$. Therefore, the central image is in perfect focus at the exit slit.

To find the variation in ρ', and thus r', under the above conditions for C_1 consider the spectral range from 0 to 2000 Å with $R = 1$ m, $1/d = 600$ lines/mm, $m = +1$, $\rho = 1/\cos \alpha_0$, $\alpha_0 = 35°15'$, and $C = 70°30'$. To cover this range the grating must be rotated through an angle $\Delta\alpha = 4°13'$ (0.07 radians). Substituting these values into (3.36), the variation in r'

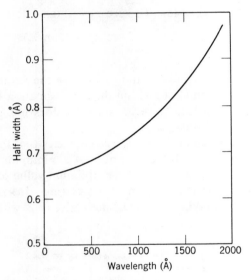

Fig. 3.20 Wavelength resolution variation for a Seya-Namioka monochromator as a function of wavelength. Straight slits 5 mm in length, 37μ wide and a 1200 line per mm grating were used.

is $\Delta r' \sim 0.007$ mm. For other spectral ranges the best values of C, ρ, and ρ' have been calculated by Greiner and Schäffer [52]. Using the same parameters as above, we find that the solution $C_3 = 0$ gives $\Delta r' \sim 1.8$ mm, hence the reason for omitting this solution.

In the Seya-Namioka mounting the constructional parameters are

$$C = 70°30';$$

$$\rho = \rho' = \frac{1}{\cos 35°15'} = 1.2245.$$

Because these parameters are independent of R and d, any grating can be used in a given monochromator by simply changing the lengths of the entrance and exit arms such that ρ and ρ' stay constant, a higher resolution being achieved for increasing R and decreasing d. Figure 3.20 shows the variation in widths of the spectral lines at half their maximum intensities as a function of wavelength for $R = \frac{1}{2}$ m and $1/d = 1200$ lines per mm.

The small amount of defocusing caused by the variation in $r' \sim 0.007$ mm and for small deviations in the parameters ρ, ρ', and C is negligible compared to the higher order aberrations in the grating equation, which are zero in the Rowland and Wadsworth mounting of the grating but *not* zero in the Seya-Namioka mounting. In fact, the astigmatic image is slightly curved and is the major cause for poor resolution when long, straight exit slits are used. The use of slightly curved exit slits greatly improves the resolution. The curvature of the spectral image is caused primarily by the aberration called *astigmatic curvature* discussed in Chapter 2. The radius of curvature of the exit slit should be made equal to R/Ψ (see (2.28)). For the Seya mounting the curvature of the spectral lines remains nearly constant between 0 and 2000 Å. Thus the radius of the exit slit is given by

$$r = 0.577R. \tag{3.48}$$

The center of curvature lies between the entrance and exit slits. Bath and Brehm [53] have obtained a wavelength resolution of 0.2 Å using a 1 m grating with 1440 lines per mm and curved slits 2 cm long and 20 μ wide. The author has obtained a resolution of 0.6 Å using a $\frac{1}{2}$ m grating with 1200 lines per mm with a curved exit slit ($r = 28.5$ cm) 1.3 cm high by 25 μ wide. The astigmatism in the Seya mounting is approximately

$$z = l + \tfrac{2}{3}L, \tag{3.49}$$

where l is the height of the entrance slit and L is the length of the grating rulings illuminated. Equation 3.49 is found to hold in practice. This astigmatism can be greatly reduced by using a toroidal grating with radii of ratio $R_v/R_h \sim 0.67$ (see Sect. 2.6).

The theoretical resolving power of the monochromator is a maximum for a grating width of approximately 2.8 cm using a 1 m grating with 600 lines per mm. For wider gratings, the resolution rapidly decreases. The intensity, on the other hand, increases with grating width until a width of approximately 4 cm is reached, at which point the intensity approaches a constant value. Thus the width of the grating must be chosen to optimize the resolution or the intensity.

The best scanning mechanism for a Seya monochromator is the sine drive, which produces a linear wavelength scale as the grating is rotated. For a simple rotation of the grating through an angle θ, the grating equation can be expressed as $m\lambda = 2d \cos \varphi \sin \theta$, where φ is equal to $\frac{1}{2}(\alpha - \beta)$ and θ is equal to $\frac{1}{2}(\alpha + \beta)$. When the exit and entrance slits are fixed, φ is constant. Therefore the wavelength appearing at the exit slit is proportional to the sine of the angle of rotation. A drive which provides a linear displacement proportional to sin θ is called a sine drive [54,55].

Miyake and Katayama [56–58] have discussed the theory for off-Rowland circle mountings more generally and have derived all the preceding mounts as special cases.

Plane Grating Mountings

Czerny-Turner and Ebert-Fastie Mounts.   The Ebert-Fastie mounting [19–21] has been described under spectrographs in Sect. 3.1. However, its

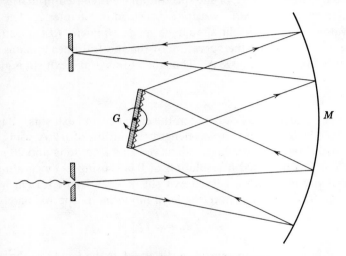

Fig. 3.21   Ebert-Fastie monochromator.

major advantage as described by Fastie is in its use as a scanning spectrometer. This instrument is basically a special case of the more general mounting described by Czerny and Turner [22]. Figure 3.21 shows the optical system of the single mirror as used in the Ebert-Fastie monochromator while Fig. 3.22 shows the two-mirror system used by Czerny and Turner. The monochromatic action is achieved by a simple rotation of the plane grating about its center.

In both systems the off-axis aberrations produced by the reflection from the first concave mirror are cancelled by the reflection from the second concave mirror. In fact, it appears that the limit of resolution is determined by imperfections in the grating and not by aberrations of the optical system. To obtain maximum resolution from these systems Fastie has shown that curved exit and entrance slits should be used, the radius of curvature being equal to the separation distance of the slits from the optical axis of the system. This circle is called the Ebert circle. Thus, the ability to use very long slits without imparing the resolution makes this mounting particularly suitable for observing extended light sources.

The three reflections required in a plane grating mounting are extremely troublesome in the vacuum ultraviolet region since the reflecting power of most materials decreases rapidly below 2000 Å. However, the combination of aluminum and magnesium fluoride coatings has proved to be an excellent reflector for radiation down to about 1200 Å [59,60]. With such high reflecting coatings, the Czerny-Turner and Fastie-Ebert monochromators probably have the most desirable features when working above 1200 Å,

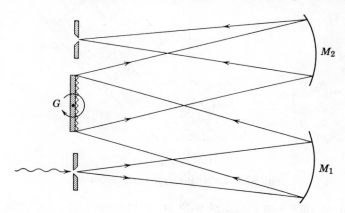

Fig. 3.22   Czerny-Turner monochromator.

namely, resolution, high light-gathering power, simple scanning mechanism, and the advantages of fixed exit and entrance slits with no deviation in the direction of the exit beam. A rather complete description of the mono-chromator has been given by Fastie [61].

Collimator Mounting. With collimators, instead of the conventional exit and entrance slits, and with a plane diffraction grating, it is possible to construct a very simple monochromator of moderate resolution. Mono-chromators of this type have often been used in the x-ray region. Until recently, however, none had been used in the vacuum uv region [62]. Figure 3.23 shows the arrangement of the components for the type of collimating monochromator constructed by Bedo and Hinteregger [62]. The wavelengths can be scanned by a simple rotation of the plane grating about an axis in the surface of the grating parallel to the rulings. The collimator is constructed using a series of grids spaced appropriately to achieve maximum collimation. Collimation can also be applied to the incident beam if it is not sufficiently parallel. Each grid consists of a series of slits etched by an electroforming process. The arrangement of

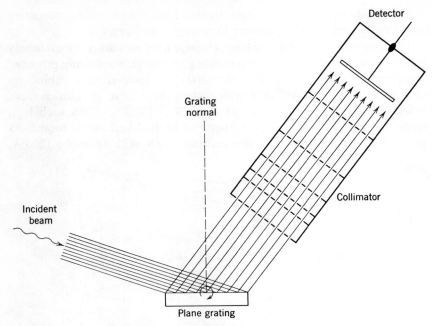

Fig. 3.23   Collimator-type monochromator.

the grids is shown in Fig. 3.24 with an acceptance angle $\Delta\alpha$. The use of grids, rather than solid parallel plates of the Soller type, improves the collimation since it eliminates diverging beams that may be transmitted by a series of internal reflections. Bedo and Hinteregger have obtained a wavelength resolution of approximately 5 Å at 584 Å when the angular divergence of the incident beam was 30' and the acceptance angle of the collimator was 10'. The grating was ruled with 1800 lines per mm and was used at an angle of incidence of 86°.

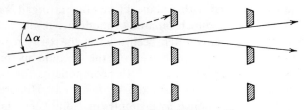

Fig. 3.24   Typical collimating grids.

Fresnel Zone Plate

The Fresnel zone plate has never been used as a practical spectroscopic instrument. It does, however, possess some optical properties that would be of practical value in the extreme ultraviolet.

The zone plate is an array of concentric circles whose radii are approximately proportional to the square roots of the consecutive integers 1, 2, 3, . . . , while the area between consecutive circles are made alternatively opaque and transparent to the incident radiation. The principle of the zone plate is described in most optical textbooks [63], while the theory has been discussed recently by Myers [64] and Kamiya [65].

Briefly, the principle of the zone plate is shown in Fig. 3.25 for parallel light. The radii of the circles are given by $r_1, r_2, . . . $, etc., and are constructed so that the distance from the circumference of a given circle to the

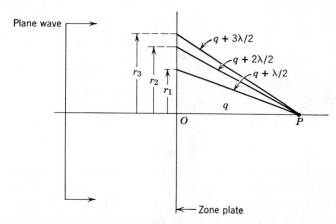

Fig. 3.25   Construction of half-period zones for a plane wave.

point P differs from that of a consecutive circle by $\lambda/2$. Therefore, if every alternate zone is made opaque to the incident radiation, then the radiation penetrating the open apertures and directed towards P will on the average have pathlengths differing by $n\lambda$, $n = 1, 2, 3, . . . $, etc., and the light intensity will be reinforced at P. It makes no difference whether the central zone is opaque or transparent: the focusing properties of the zone plate are the same. The optical properties of the zone plate are as follows:

RADIUS r_n OF nth RING

$$r_n{}^2 = fn\lambda + \frac{n^2\lambda^2}{4}, \qquad n = 1, 2, 3, . . . , \qquad (3.50)$$

where f is the focal length of the zone plate for a wavelength λ;

FOCAL LENGTH $$f = \frac{r_1{}^2}{m\lambda}, \qquad m = 1, 3, 5, . . . , \qquad (3.51)$$

when $\lambda/4 \ll f$, and where m is the order number. The primary image is given when $m = 1$ and is the most intense image. The images appearing for $m = 3, 5, \ldots$ are less intense and are approximately in the ratio $1 : m$. A second order image $m = 2$ may be observed but is due to any inaccuracy in the construction of the zone plate.

FOCAL EQUATION
$$\frac{1}{p} + \frac{1}{q} = \frac{1}{f},$$
(3.52)

where p and q are, respectively, the object and image distances.

MINIMUM ANGLE OF RESOLUTION

$$\theta = \frac{1.22\lambda}{D},$$
(3.53)

where D is the diameter of the zone plate.

DISPERSION
$$\frac{dq}{d\lambda} = \frac{mr_1^2}{(m\lambda - r_1^2/p)^2} \qquad (m = 1, 3, 5, \ldots)$$
(3.54)

$$= \frac{r_1^2}{m\lambda^2}, \qquad \text{for} \quad p \rightarrow \infty.$$
(3.55)

For the zone plate to be of practical use in the vacuum ultraviolet, the transparent zones must be completely open. Baez [66] has succeeded in constructing such a zone plate with 38 zones. The opaque elements were thin concentric bands of gold made self-supporting by the use of thin radial struts, as shown in Fig. 3.26. The central circle had a diameter of 0.0426 cm, while the outer circle and diameter of the zone plate was 0.2596 cm. The error in the circle diameters was ±0.0002 cm.

Baez and Myers have suggested the use of zone plates as a focusing device for extreme uv and x-rays. Equation 3.52 above shows that the zone plate has focal properties which are similar to a simple lens. Using the open construction type zone plate, Baez has photographed the image of a mesh with 4 lines per mm illuminated with light of wavelength 6700 Å, 4538 Å, and 2537 Å. Figure 3.27 shows the images normalized to the same magnification. The resolution of the images clearly increases as the wavelength decreases. This was predicted by (3.53). It would be expected, therefore, that excellent resolution would be obtained at shorter wavelengths. Baez has shown that the zone plate, compared to a pinhole camera, has higher resolution and greater light-gathering power. In fact, the optimum diameter of a pinhole is given by [67]

$$D = 2\sqrt{\frac{0.9pq\lambda}{p + q}},$$
(3.56)

Fig. 3.26 Photograph of a self-supported gold zone plate. The diameter of the outer circle is 0.26 cm while the diameter of the central circle is 0.043 cm. The thickness of gold is estimated as 10 microns. The white bands represent the completely open structure (courtesy A. V. Baez [66]).

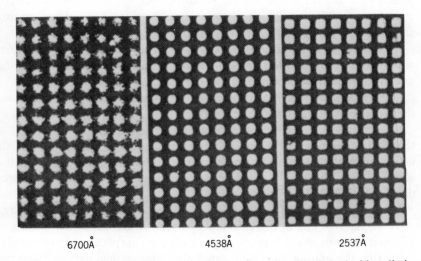

6700Å   4538Å   2537Å

Fig. 3.27 Zone plate images of a mesh with 4 lines per mm illuminated with radiation of wavelength 6700, 4538, and 2537 Å (courtesy A. V. Baez [66]).

which from (3.52) becomes

$$D = 2\sqrt{0.9f\lambda}.$$ (3.57)

This is practically the diameter of the innermost circle of a zone plate.

From (3.51) we see that the zone plate is very highly chromatic. For a radius r_1 equal to 0.02 cm, the focal length f is equal to 400 cm at 100 Å and is equal to 80 cm at 500 Å. It is this fact that led Kamiya to suggest that the zone plate may be usable as a monochromator in the extreme uv. The dispersion of a zone plate for parallel light is given by (3.55). For r_1 equal to 0.02 cm, and $m = 1$, the reciprocal dispersion is 0.025 Å per mm at 100 Å, 0.625 Å per mm at 500 Å, and 2.5 Å per mm at 1000 Å. Although the reciprocal dispersion is very small, the practical difficulties of utilizing the zone plate to obtain band passes of a few angstroms appear formidable. However, for specific purposes, it does seem as if the zone plate is a practical tool. For example, it could be used as a pre-disperser with a few hundred angstrom band pass. As a narrow band filter, it should be very good at separating lines in a sparsely distributed emission line source such as the 584 Å He I line in a helium DC glow discharge or in separating the 584 He I and 304 He II lines in the solar spectrum.

3.3 FOCUSING

The general principles of focusing a concave grating spectrograph or monochromator are as follows: The centers of the slits, grating, and plateholder should all lie on the circumference of the Rowland circle. Furthermore, the lengths of the slits and the ruled lines of the grating should be perpendicular to the Rowland plane.

To achieve these conditions the slit assembly is generally manufactured with two degrees of motion: rotation about the axis joining slit and grating and linear motion along this axis. The grating should have three degrees of rotation about mutually perpendicular axes whose origin coincides with the center of the grating surface.

The entrance slit is adjusted parallel to the vertical by means of a theodelite or similar instrument. The grating is then rotated about an axis coincident with the grating normal to align the rulings with the length of the entrance slit. The quality of the image can be determined by visual inspection in the initial focusing procedure.

To position the Rowland circle in a horizontal plane, the central image or some suitable line in the visible is observed as the tilt of the grating is adjusted until the image is properly positioned on the plateholder or exit slit.

Finally, the grating must be rotated about its vertical axis to bring the grating normal into the correct position such that the entrance and exit slits will lie on the circumference of the Rowland circle. If the center of the Rowland circle has been located previously, the positioning of the grating normal is straightforward since it must pass through the center of the Rowland circle. The procedure used by McPherson is to stretch a thin wire vertically over the grating mount, passing over its center, and then to locate a conical bob at the center of the Rowland circle. The point of the bob is lined up visually with the vertical wire, while the grating is rotated about its vertical axis until the image of the bob is also in line with the wire, bob, and eye.

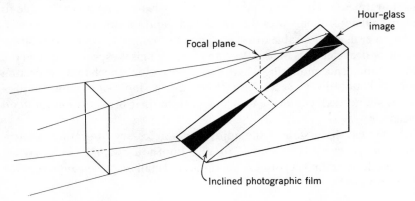

Fig. 3.28 Hour-glass image for locating the focal plane.

The plateholder is machined to follow the curvature of the Rowland circle to as high a degree as possible. It can be helpful if the plate-holder is pivoted about the point of intersection of the grating normal with the Rowland circle since the plateholder can then be rotated about this point until it coincides with the Rowland circle. This point can be located rather precisely by using the Foucault knife-edge test. The principle of this test is that if the image of a small illuminated hole is observed at the focal plane, then the grating is seen to be filled with light; and if a knife-edge is passed through the focal plane, the entire grating should suddenly and uniformly become dark. When the knife edge is moved across the wedge of light a short distance inside the focus, its shadow, as it appears on the grating, moves in the same direction as the knife-edge; if it is placed outside the focus, the shadow moves in the opposite direction. The knife-edge test is remarkably sensitive and can be used to locate the position of the focal plane to within approximately 0.1 mm.

Where tests must be made on the nonvisible lines, the "hour-glass" technique can be used. This method is illustrated in Fig. 3.28. A wedge is made with the sloping face inclined at a few degrees to the horizontal.

Then by mounting a photographic film onto the inclined face, the wedge is inserted into the diffracted beam. An image shaped like an hour-glass results. The narrow neck of the image is the position of the focal plane. A datum line can be engraved on the wedge indicating the plane of the plateholder. The deviation between the datum line and the narrowest part of the image indicates the amount and direction of the correction. Using this technique, Shenstone [68] has observed in his spectrograph that while the first, second, and fourth order spectra were sharply focused, the third order was not. This anomaly is not well understood but may be common to most gratings. Thus when optimum focusing is required it is advisable to make final adjustments with the wavelengths and spectral order of interest.

In general, it is desirable to use precision machining to locate the position of the optical elements on the Rowland circle. This is especially true in the grazing incidence mounting where the construction is such that the grating normal is not always accessible. When this is done, only minor adjustments are necessary to bring the elements into their best position for good focus.

A detailed discussion of the adjustments necessary in grating spectrographs is given by Harrison, Lord, and Loofbourow [69], while Sawyer [70] and MacAdam [71] discuss the optics of small displacements from the Rowland circle.

REFERENCES

[1] A. Eagle, *Astrophys. J.* **31,** 120 (1910).
[2] A. Eagle, *Proc. Phys. Soc.* **23,** 233 (1911).
[3] P. G. Wilkinson, *J. Mol. Spectroscopy* **1,** 288 (1957).
[4] T. Namioka, *J. Opt. Soc. Am.* **49,** 460 (1959).
[5] A. E. Douglas, *J. Opt. Soc. Am.* **49,** 1132 (1959).
[6] A. E. Douglas and J. G. Potter, *Appl. Optics* **1,** 727 (1962).
[7] F. Tyrén, *Zeits. f. Physik* **111,** 314 (1938).
[8] H. A. Kirkpatrick, *J. Quant. Spect. Rad. Transfer* **2,** 715 (1962).
[9] A. H. Gabriel, J. R. Swain, and W. A. Waller, *J. Sci. Instr.* **42,** 94 (1965).
[10] G. A. Sawyer, A. J. Bearden, I. Henins, F. C. Jahoda, and F. L. Ribe, *Phys. Rev.* **131,** 1891 (1963).
[11] I. W. Ruderman, K. J. Ness, and J. C. Lindsay, *Appl. Phys. Letters* **7,** 17 (1965).
[12] A. A. Michelson, *Astrophysics J.* **8,** 37 (1898); *Proc. Am. Acad.* **35,** 111 (1899).
[13] From the design of Michelson, Williams produced the first successful reflection echelon. W. E. Williams, *Nature* **127,** 816 (1931); *Proc. Phys. Soc.* **45,** 699 (1933).
[14] R. W. Wood, *J. Opt. Soc. Am.* **37,** 733 (1947).
[15] G. R. Harrison, *J. Opt. Soc. Am.* **39,** 522 (1949).
[16] G. R. Harrison, J. E. Archer, J. Camus, *J. Opt. Soc. Am.* **42,** 706 (1952).
[17] G. R. Harrison and G. W. Stroke, *J. Opt. Soc. Am.* **50,** 1153 (1960).
[18] G. W. Stroke and H. H. Stroke, *J. Opt. Soc. Am.* **53,** 333 (1963).

[19] H. Ebert, *Wied. Ann.* **38,** 489 (1889).

[20] W. G. Fastie, *J. Opt. Soc. Am.* **42,** 641 (1952).

[21] W. G. Fastie, *J. Opt. Soc. Am.* **42,** 647 (1952).

[22] M. Czerny and A. F. Turner, *Z. Physik.* **61,** 792 (1930).

[23] F. L. O. Wadsworth, *Astrophys. J.* **3,** 54 (1896).

[24] W. F. Meggers and K. Burns, *Bur. Standards Sci. Papers* **18,** 185 (1922).

[25] R. Tousey, *Space Science Reviews* **2,** 3 (1963).

[26] R. F. Baker, *J. Opt. Soc. Am.* **28,** 55 (1938).

[27] E. R. Piore, G. G. Harvey, E. M. György, and R. H. Kingston, *Rev. Sci. Instr.* **23,** 8 (1952).

[28] D. O. Landon, *Appl. Optics,* **3,** 115 (1964).

[29] H. E. Hinteregger, *Astrophys. J.* **132,** 801 (1960).

[30] H. E. Hinteregger, *Space Astrophysics,* ed. W. Liller (McGraw-Hill, New York, 1961), pp. 35–73.

[31] McPherson Instrument Corp., Acton, Mass.

[32] M. Salle and B. Vodar, *Compt. Rend.* **230,** 380 (1950).

[33] R. Tousey, F. S. Johnson, J. Richardson, and N. Toran, *J. Opt. Soc. Am.* **41,** 696 (1951).

[34] Y. Fujioka and R. Ito, *Sci. of Light* (Tokyo) **1,** 1 (1951).

[35] A 40-cm radius mounted monochromator is commercially available from the Tropel Co.

[36] F. J. P. Clarke and W. R. S. Garton, *J. Sci. Instr.* **36,** 403 (1959).

[37] B. Vodar, *Rev. Optique* **21,** 97 (1942).

[38] S. Robin, *J. Phys. Radium* **14,** 551 (1953).

[39] McPherson 1 M monochromator, Model 225.

[40] R. Tousey, An Apparatus for Determining the Optical Constants of Solids in the Extreme Ultraviolet with Application to Fluorite at 1216 Å (Thesis, Harvard University, June 1933).

[41] R. Ditchburn, University of Reading, England, private communication.

[42] M. Pouey and J. Romand, *Revue d'Optique* **44,** 445 (1965).

[43] T. Namioka, *J. Opt. Soc. Am.* **49,** 961 (1959).

[44] E. J. Bair, P. C. Cross, T. L. Dawson, A. E. Wilson, and J. H. Wise, *J. Opt. Soc. Am.* **43,** 681 (1953).

[45] P. D. Johnson, *Rev. Sci. Instr.* **28,** 833 (1957).

[46] R. Onaka, *Sci. Light* (Tokyo) **7,** 23 (1958).

[47] W. W. Parkinson, Jr., and F. E. Williams, *J. Opt. Soc. Am.* **39,** 705 (1949).

[48] G. L. Weissler, *Handbuch der Physik* (Springer-Verlag, Berlin, 1956), Vol. XXI.

[49] M. Seya, *Sci. Light* **2,** 8 (1952).

[50] T. Namioka, *Sci. Light* **3,** 15 (1954).

[51] T. Namioka, *J. Opt. Soc. Am.* **49,** 951 (1959).

[52] H. Greiner and E. Schäffer, *Optik* **14,** 263 (1957); **15,** 51 (1958).

[53] K. L. Bath and B. Brehm, *Z. Angew. Physik* **19,** 39 (1965).

[54] R. M. Badger, *Rev. Sci. Instr.* **19,** 861 (1948).

[55] K. Andō, *Sci. Light* **13,** 45 (1964).

[56] K. P. Miyake, *Sci. Light* **8,** 39 (1959).

[57] K. P. Miyake and T. Katayama, *Sci. Light* **11,** 1 and 10 (1962).

[58] T. Katayama, *Sci. Light* **11,** 21 (1962).

[59] G. Hass and R. Tousey, *J. Opt. Soc. Am.* **49,** 593 (1959).

[60] P. H. Berning, G. Hass, and R. P. Madden, *J. Opt. Soc. Am.* **50,** 586 (1960).

[61] W. G. Fastie, H. M. Crosswhite, and P. Gloersen, *J. Opt. Soc. Am.* **48,** 106 (1958).

[62] D. E. Bedo and H. Hinteregger, *J. Appl. Physics* (Japan) Supplement 1, **4,** 473 (1965).

[63] R. W. Ditchburn, *Light* (Interscience, New York, 1953); F. A. Jenkins and H. E. White, "Fundamentals of Physical Optics" (McGraw-Hill, New York, 1937); and R. W. Wood, *Physical Optics* (Macmillan, New York, 1914).

[64] O. E. Myers, Jr., *Am. J. Phys.* **19,** 359 (1951).

[65] K. Kamiya, *Sci. Light* **12,** 35 (1963).

[66] A. V. Baez, *J. Opt. Soc. Am.* **51,** 405 (1961).

[67] J. E. Mack and M. J. Martin, *The Photographic Process* (McGraw-Hill, New York, (1939).

[68] A. G. Shenstone, *J. Opt. Soc. Am.* **53,** 1253 (1963).

[69] G. R. Harrison, R. C. Lord, and J. R. Loofbourow, *Practical Spectroscopy* (Prentice-Hall, Englewood Cliffs, N.J., 1948).

[70] R. A. Sawyer, *Experimental Spectroscopy* (Dover, New York, 1963), 3rd ed.

[71] D. L. MacAdam, *J. Opt. Soc. Am.* **23,** 178 (1933).

4

Vacuum Techniques

The degree of vacuum necessary in a spectrograph to prevent any appreciable attenuation of the radiation from the light source depends on the wavelength and the pathlength from the entrance slit to the plateholder via the diffraction grating. The amount of attenuation of the radiation is given by the Lambert-Beer Law, namely,

$$I = I_0 e^{-\sigma n L}, \tag{4.1}$$

where I_0 is the intensity of the unattenuated beam while I is the intensity reaching the plateholder or exit slit; the pathlength traveled by the radiation is L; the number of molecules in the spectrograph is represented by n per cubic centimeter, and the absorption cross section of the molecules (air) is represented by σ. The absorption cross section of a gas is a function of wavelength and in general is greater at wavelengths shorter than the first ionization potential of the gas. Since the first ionization potential of most gases lie below 1100 Å, and no window materials exist below 1040 Å, the minimum vacuum requirements are divided naturally into two regions above and below 1100 Å. The region below 1100 Å not only requires a better ultimate vacuum than the region above 1100 Å, but it also requires higher pumping speeds to remove the gas, which continually diffuses in through the entrance slit from the light source. Pressures of 10^{-4} to 10^{-6} torr are quite adequate for most vacuum spectrographs and monochromators.

The choice of vacuum pumps, both mechanical and vapor types, is determined mainly by the following: the desired pump down time from atmospheric pressure to an operational pressure; the desired ultimate vacuum; and the pumping speed necessary to handle gas leakage from the light source, windowless absorption cells, and any spurious leaks in the system.

The time taken to exhaust a volume V, with no leaks and with no appreciable outgassing, from a pressure P_1 to P_2 is given by

$$t = 2.3 \frac{V}{S} \log \frac{P_1}{P_2}, \tag{4.2}$$

85

where S is the speed of the pump expressed in units of volume per unit of time, typically in liters per second or cubic feet per minute (CFM). To convert one into the other, divide CFM by 2.12 to give liters per sec. If the pumping speed varies over the pressure range, it is necessary to apply (4.2) to successive small intervals of pressure and add up the several incremental intervals. Figures 4.1 (a), (b) and (c) show typical curves of pumping speed as a function of pressure for several different commercial mechanical and vapor pumps. It should be noted, however, that the use of a cooled baffle decreases the pumping speed of the system by about a factor of 2 below 10^{-3} torr. If the coolant is liquid nitrogen, the ultimate vacuum reached can be more than an order of magnitude lower.

The use of a cooled baffle between the final pumping stage and the spectrograph is essential to minimize the back diffusion of the pump oil, especially that of the forepump whose oil has a vapor pressure in the vicinity of 10^{-4} torr. The presence of oil in the spectrograph is particularly

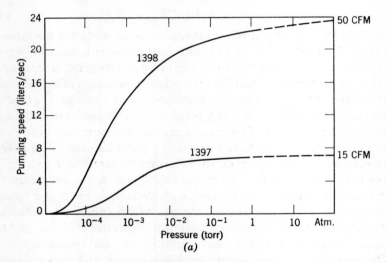

Fig. 4.1 (a) Pumping speeds of Welch mechanical pumps. (b) Pumping speeds of Heraeus Roots oil-less mechanical pumps. These pumps must be backed by a forepump with the following recommended characteristics:

	R-152	RG-350	RG-1000
forepump speed (CFM)	15	47	130
max. forepressure (torr)	20	50	50

(c) Pumping speeds of various diffusion pumps. The PMC-720 and 1440 pumps are produced by Consolidated Vacuum Corporation with 5″ and 7″ I. D. flanges, respectively, while the B-4 and B-6 booster pumps with 4″ and 6″ I. D. flanges, respectively, are produced by the NRC Equipment Corporation. The maximum forepressure for the booster pumps at maximum throughout is 0.8 torr and for the PMC series, it is 0.5 torr.

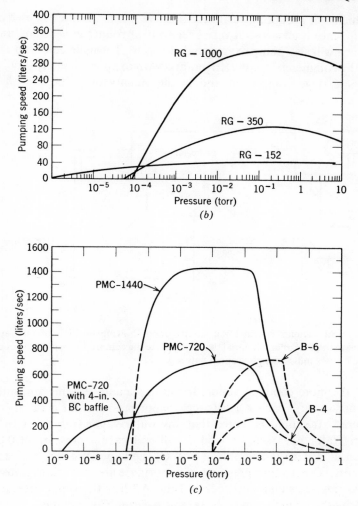

Fig. 4.1   (*continued*)

harmful to the reflecting power of the grating. Any oil on the grating surface is soon decomposed by the action of the incident ultraviolet radiation, and the resulting carbon compound coating adheres to the grating surface and cannot generally be cleaned. This coating has a poor reflectance compared to that of the standard high reflectance coatings. Over and above the bad effects oil has on the diffraction grating, it provides a contaminant in various windowless experiments. Since the forepump oil is the worst offender, it is desirable to use an oil trap between the mechanical pump and diffusion pump. Figure 4.2 shows a block diagram of a system which will produce a relatively clean vacuum. Zeolite* is used in the oil trap since it adsorbs oil very effectively at room temperature, and when cooled, will also pump oxygen and nitrogen. The main cold trap can be cooled by commercial freon refrigerants down to approximately $-50°C$ with a single stage compressor and in the vicinity of $-100°C$ with a two-stage compressor.

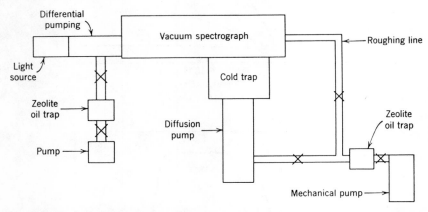

Fig. 4.2 Typical vacuum system for a spectrograph showing position of oil traps. Molecular sieve traps are manufactured that can replace the cold trap above the diffusion pump. The crosses indicate high vacuum valves.

Two of the more common vapors found in vacuum systems are water and mercury vapor. Table 4.1 lists their vapor pressures for a variety of temperatures. It should be noted that any water vapor frozen out by a freon refrigerator operating at $-50°C$ will still have a pressure of 0.03 torr. Even at dry-ice temperature $(-78°C)$ the vapor pressure of ice is 5×10^{-4} torr. Other vapor pressures of importance are the various greases used for O-ring seals and stop cocks. Table 4.2 lists the vapor pressure, as quoted by the manufacturers, of several Apiezon Compounds.

The total pumping speed S_T of the pump and connecting tubing can be determined from the relation

$$\frac{1}{S_T} = \frac{1}{S_p} + \frac{1}{C}, \tag{4.3}$$

where S_p is the speed of the pump and C is the pumping speed or conductance of the connecting tubing. The conductance of the tubing depends on whether the gas flow is viscous or molecular flow. That is, whether the molecules make more frequent collisions with each other or

* Molecular Sieves Department, Linde Company, Division of Union Carbide Corp.

Table 4.1 Vapor Pressure of Ice and Mercury[a]

Ice		Mercury	
$t\,^\circ C$	P (torr)	$t\,^\circ C$	P (torr)
0	4.58	40	6.1×10^{-3}
-10	1.96	30	3.1×10^{-3}
-20	7.8×10^{-1}	25	2.0×10^{-3}
-30	2.9×10^{-1}	20	1.4×10^{-3}
-40	1.0×10^{-1}	10	5.6×10^{-4}
-50	3.0×10^{-2}	0	2.2×10^{-4}
-60	8.0×10^{-3}	-10	7.9×10^{-5}
-70	2.0×10^{-3}	-20	2.6×10^{-5}
-78.5	5.0×10^{-4}	-30	7.9×10^{-6}
-80	4.1×10^{-4}	-38.9	2.5×10^{-6}
-90	7.5×10^{-5}	-45	1.1×10^{-6}
-100	1.1×10^{-5}	-50	4.9×10^{-7}
-110	1.3×10^{-6}	-60	9.9×10^{-8}
-120	1.1×10^{-7}	-70	1.7×10^{-8}
-130	7.0×10^{-9}	-78.5	3.1×10^{-9}
-140	2.9×10^{-10}	-80	2.4×10^{-9}
-150	7.4×10^{-15}	-100	2.4×10^{-11}
-183	1.4×10^{-22}	-183	3.5×10^{-32}

[a] Adapted from S. Dushman, *Scientific Foundations of Vacuum Techniques*, J. M. Lafferty, Editor (Wiley, New York) 2nd ed.

Table 4.2 Vapor Pressure of Apiezon Compounds[a]

Compound	Vapor Pressure After Evolution of Dissolved Air at 20°C (Torr)	Melting Point (°C)
Apiezon grease L	10^{-10} to 10^{-11}	47
Apiezon grease M	10^{-7} to 10^{-8}	44
Apiezon grease N	10^{-8} to 10^{-9}	43
Apiezon grease T	10^{-8} (approx)	125
Sealing compound Q	10^{-4}	45

[a] Courtesy James G. Biddle Co., Pa.

with the walls of the tubing. There is no sharp dividing line between the two types of flow, however, the following criteria can be used.

$$\text{viscous flow} \quad L < \frac{d}{100},$$

$$\text{molecular flow} \quad L > d,$$

where L is the mean free path of the molecules evaluated at the average pressure in the tubing and d is the diameter of the tubing (for a rectangular slit d is to be associated with the slit width).

Table 4.3 Diameters and Molecular
Weights of Some Gases[a]

Gas	Molecular Weight	Molecular Diameter $(10^{-8}$ cm)
He	4.00	2.0
Ne	20.18	2.6
Ar	39.94	2.9
Kr	83.80	3.1
Xe	131.30	3.4
H_2	2.02	2.4
O_2	32.00	3.0
N_2	28.02	3.2
CO	28.01	3.2
CO_2	44.01	3.3

[a] *Handbook of Chemistry and Physics*, (Chemical Rubber Publ. Co., Cleveland, Ohio, 1958) 40th ed.

The range lying between these values is known as the transition range.

From kinetic theory, the mean free path is given by

$$L = \frac{1}{n\pi D^2 \sqrt{2}}, \tag{4.4}$$

where D is the molecular diameter in centimeters and n is the number of molecules per cc given by $n = (P_\mu/T)(9.7 \times 10^{15})$ when the pressure P_μ is measured in microns and the temperature T is measured in °K. Table 4.3 lists the diameters and molecular weights of several molecules. For air at room temperature,

$$L \approx \frac{5}{P_\mu} \quad \text{cm.} \tag{4.5}$$

In the relatively high pressure range characteristic of the operation of single stage mechanical vacuum pumps, the flow is generally viscous. When the length l of the connecting tubing is much greater than its diameter d, then the conductance of the tubing for air at 25°C is given by

$$C = 0.18 \frac{d^4}{l} P_\mu \qquad \text{liters/sec,} \qquad (4.6)$$

where d and l are in cm and P_μ is the average pressure measured in microns. For gases other than air (4.6) must be multiplied by the ratio of the viscosity of air to that of the gas concerned.

The conductance for air at 25°C in the molecular flow region for $l \gg d$ is given by

$$C = 12.2 \frac{d^3}{l} \qquad \text{liters/sec,} \qquad (4.7)$$

where d and l are measured in cm. For gases other than air, (4.7) must be multiplied by the square root of the ratio of the molecular weight for air ($M = 29$) to that of the gas concerned. Equations 4.6 and 4.7 refer to circular cross sections only.

This discussion on pumping speeds has an important application to differential pumping commonly used between the light source and the spectrograph. Thus it is of value to present the equations for the conductances most often encountered. The following equations apply strictly to molecular flow. Since viscous flow is somewhat faster, these equations represent a lower limit to the pumping speeds.

The general expression for the conductance of a canal of any cross section is given by

$$C = \frac{3.6\, A\sqrt{T/M}}{1 + (3/16)(Hl/A)} \qquad \text{liters/sec,} \qquad (4.8)$$

where A is the cross sectional area, T is the temperature in °K, M is the molecular weight, H is the perimeter of the opening, and l is the length of the canal. All units are in cm.

For a circular cross section of diameter d, using air, (4.8) becomes

$$C = 12.2 \frac{d^3}{l + \frac{4}{3}d} \qquad \text{liters/sec,} \qquad (4.9)$$

which is the same as (4.7) when $l \gg d$.

For a rectangular cross section with sides a and b, the conductance for air is

$$C = 31 \frac{a^2 b^2}{2.7ab + l(a + b)} \qquad \text{liters/sec.} \qquad (4.10)$$

When $l \sim 0$, (4.8) gives the conductance of an aperture, namely,

$$C = 11.6A \qquad \text{liters/sec.} \tag{4.11}$$

Figure 4.3 shows a schematic diagram of the light source at a pressure P_1, the differential pumping unit at a pressure P_2, and the main spectrograph at a pressure P_3. The speed of the differential pump is represented by S_p, and that of the main system as S'_p. The effective pumping speed S_T at the entrance to the differential chamber is given by (4.3), namely,

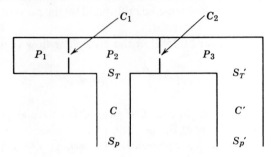

Differential pumping

Fig. 4.3 Differential pumping system. P_1, light source pressure; P_2, differential pumping unit pressure; P_3, spectrograph pressure; S_T and S'_T, the effective pumping speed of the entrance to the differential chamber and spectrograph, respectively; C and C', conductance of the connecting tubes; S_p and S'_p, pumping speeds of the differential pump and spectrograph pump, respectively.

$S_T = CS_p/(C + S_p)$. The conductance C of the connecting line is given either by (4.6) or (4.9), depending on the type of flow, whereas C' is generally represented by (4.9).

The pumping speed at any point in a vacuum system is defined as

$$S = \frac{Q}{P}, \tag{4.12}$$

where Q is the gas flow in the system and S is defined as the pumping speed at the point where the pressure P is measured. From (4.3) and (4.12) it follows that the gas flow across an aperture such as C_1 in Fig. 4.3 is

$$Q = (P_1 - P_2)C_1. \tag{4.13}$$

Since the gas flow Q is a constant for the system, and if $S_T \gg C_2$, most of the gas will flow through the differential pump and $Q \simeq P_2 S_T$. Therefore for the differential pumping system shown in Fig. 4.3

$$P_2 = P_1 \frac{C_1}{C_1 + S_T}, \tag{4.14}$$

and

$$P_3 = P_2 \frac{C_2}{C_2 + S'_T}.$$ (4.15)

One of the apertures, of course, will be the rectangular entrance slit of the spectrograph. In order to get maximum light intensity into the spectrograph, it is desirable that the other aperture also be rectangular. The conductances C_1 and C_2 will then be determined by (4.10) or (4.11). From (4.10), it can be seen that the conductance of the aperture will be decreased by giving some depth l to the opening, thereby increasing the differential pressure.

The pumping speeds of various mechanical and diffusion pumps suitable for differential pumping are shown in Fig. 4.1a, b, c.

GENERAL REFERENCES

[1] S. Dushman and J. M. Lafferty, ed. *Scientific Foundations of Vacuum Technique* (Wiley, New York, 1962) 2nd ed.
[2] A. E. Barrington, *High Vacuum Engineering* (Prentice-Hall, Englewood Cliffs, N.J., 1963).

5

Light Sources

5.1 INTRODUCTION

Because of the great variety of experiments in vacuum ultraviolet radiation physics, no single light source would satisfy all experimental requirements. In one case, a continuous spectrum may be desired; in another, a line source may be necessary. In other cases, intense microsecond pulses of radiation or a dc source may be required. Further, with the exception of synchrotron radiation, no single light source exists which would cover the wavelength range 2 to 2000 Å.

The choice of source depends on the application. Although continuum sources are generally more desirable for providing information at all wavelengths for absorption studies, much valuable work can be done with a line spectrum, especially in the reproduction of important solar emission lines such as the 1215.7 and 1025.7 Å H I lines and the 584.4 and 304 Å He lines. A line spectrum is often more intense than a continuum. It is not always necessary to have a high-resolution spectrometer to produce highly monochromatic lines, since the separation of the lines may be one or two angstroms, whereas the width of a line is simply that produced by the source. The widths of most lines lie between 0.1 and 0.001 Å. For example, a glow discharge in helium produces the intense 584 Å He I line completely isolated from any neighboring lines, and, as a result, a wavelength resolution of about 47 Å will produce pure 584 Å radiation (the second member of the He I series lies at 537.1 Å). Although the two forbidden helium lines at 600.5 and 591.4 Å are present, they are approximately 2000 times less intense than the 584 Å line (see Fig. 5.43). It is easier to assess the amount of scattered radiation present in a line spectrum than in a continuum.

A line spectrum is produced by electronic transitions within excited atoms and molecules. Continuous radiation can be produced both by the interaction of electrons with atoms or molecules and by the acceleration (synchrotron radiation) or deceleration (bremsstrahlung radiation) of free electrons. In the first, continua are produced by two mechanisms: (a) by

94

the recombination of ions and electrons, although recombination radiation from molecular ions has not yet been observed, and (b) by a transition from a stable excited electronic state of the molecule to a lower repulsive state. The potential energy curve diagram for this case is shown in Fig. 5.1 and pertains to molecular hydrogen. According to the Franck-Condon principle, transitions take place from the various vibrational levels of the excited electronic state to points on the lower potential curve vertically below. That is, if the first five vibrational levels of the $^3\Sigma_g{}^+$ state are excited, as shown in Fig. 5.1, the most probable transitions will take place

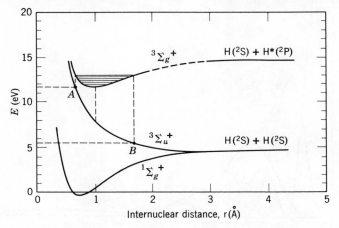

Fig. 5.1 Potential energy curves for H_2 illustrating the transitions that produce the hydrogen continuum.

between the points A and B on the repulsive $^3\Sigma_u{}^+$ curve. These transitions correspond to a continuum extending over an energy range ΔE. The continua produced by the rare gases are also caused by transitions similar to that described above. The repulsive state in this case is the ground state of the rare gas molecule.

The construction of various types of light sources and the methods for their excitation are discussed in the following sections. There are many light source designs which produce similar spectra; some more intense than others, some with slightly different spectral distributions. This is particularly true for the production of line spectra.

5.2 CONTINUUM SOURCES

Hydrogen Glow Discharge

The mechanism for the production of the H_2 continuum has been described in the last section as being due to transitions from the stable

$^3\Sigma_g{}^+$ upper state to the repulsive $^3\Sigma_u{}^+$ lower state. The continuum extends from about 1600 to 5000 Å. The great extent of the continuum is caused by the very steep potential curve of the lower state, and the number of vibrational states populated in the upper state. The intensity distribution of the H_2 continuum has been analyzed theoretically between 1600 and 4000 Å [1,2]. This distribution has been approximately verified experimentally by Coolidge between 2400 and 3400 Å [3]. The relative spectral

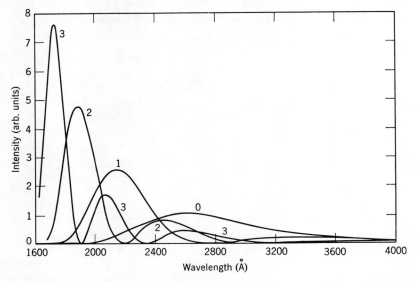

Fig. 5.2 Theoretical intensity distribution of the continuous radiation due to the $^3\Sigma_g{}^+ \to {}^3\Sigma_u{}^+$ transition from the various discrete vibrational levels of the upper $^3\Sigma_g{}^+$ state. The population of the vibrational levels are assumed to be equal (courtesy H. M. James and A. S. Coolidge [2]).

intensity distributions for transitions from each vibrational level v of the $^3\Sigma_g$ state to the continuum state are given, according to Coolidge et al. [1], by the equation

$$I_v(\lambda)\, d\lambda = CN_v\, |D_v|^2\, E^{-\frac{1}{2}}\lambda^{-6}\, d\lambda, \qquad (5.1)$$

where D_v is the transition moment, N_v is the population of the vth state, C is a constant, and E is the energy of the continuum state above the dissociation limit. Using the Franck-Condon principle, the transition moment in (5.1) is replaced by the overlap integral, and C is replaced by a different numerical constant. Figure 5.2 gives the value of the radiated intensity obtained from (5.1) plotted as a function of wavelength for each of the vibrational levels, assuming they are all equally populated. The lack of precise knowledge of the population of the various excited states makes it very difficult to compare theory with experimental results.

Figure 5.3 shows the hydrogen continuum from 1675 to 2300 Å as obtained from a dc glow discharge operated with a current of 300 mA at 200 watts. Superimposed on the continuum are the second order lines of the intense line spectrum above 1040 Å. A lithium fluoride window was used in order to remove second order lines caused by radiation of wavelengths shorter than 1040 Å, the transmission limit of lithium fluoride. Thus a smooth continuum is produced from 1675 to 2080 Å. To extend the

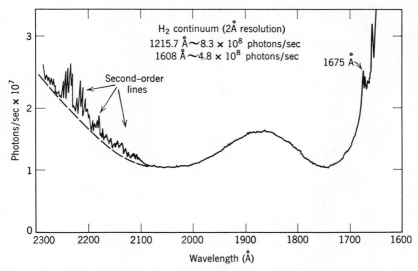

Fig. 5.3 The H_2 continuum transmitted through an LiF window. The dashed curve indicates the H_2 continuum in the absence of second-order lines (data obtained with a 1200 line per mm grating blazed at 1300 Å and coated with Al + MgF_2).

usefulness of the continuum to longer wavelengths, quartz or sapphire windows should be used. The dashed line in Fig. 5.3 represents the continuum in the absence of second order lines. Below 2000 Å, the peak intensity of the continuum lies at approximately 1860 Å. The absolute intensity of the continuum is considerably weaker than the line spectrum below 1675 Å. For comparison, the absolute intensities of the 1608 Å line and the 1215.7 Å line are shown in the figure for the particular conditions of excitation used to produce the continuum. The relative intensities of these lines and of the continuum vary with the hydrogen pressure and with the power of the discharge. Figure 5.4 shows the variation in intensity of the continuum at 1860 Å as a function of the power and current. The increase in the molecular line intensity, however, was not as rapid. The intensity of both the continuum and the molecular lines increases with increasing pressure at least over the range 0 to 2 torr.

The general shape of the continuum is characteristic of that shown in Fig. 5.3. Although this curve was not corrected for the grating efficiency, a similar curve was obtained using a 1200 line per mm grating blazed at 700 Å and coated with platinum. The ratio of peak to valley intensities was different; however, the wavelengths of the peaks and valleys agreed within about 25 Å. From the theoretical curves shown in Fig. 5.2, it

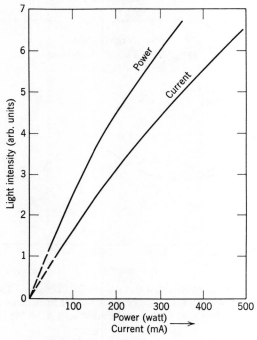

Fig. 5.4 Intensity of the H_2 continuum at 1860 Å as a function of the current and power dissipated in the light source.

appears as if the continuum at 1860 Å is produced by transitions from the second and higher vibrational levels of the $^3\Sigma_g{}^+$ state. Thus it might be expected that the intensity of the continuum would increase with increasing power and pressure. The use of platinized capillaries enhances the continuum since platinum is an excellent catalyst for the recombination of atomic hydrogen. Details of the dc glow discharge tube and other mechanisms for the excitation of the hydrogen spectrum are given in Sect. 5.3.

Rare Gas Continua

In 1930, Hopfield [4–6] discovered a continuous emission spectrum of helium extending from 600 to 1000 Å. He ascribed the continuum to a transition from the upper $^1\Sigma_u{}^+$ state to the unstable $^1\Sigma_g{}^+$ ground state of the helium molecule. This continuum has been studied extensively by Tanaka and Huffman, who have perfected the techniques necessary to produce a much more intense continuum than was originally realized by Hopfield [7–14].

The continuum is best produced by a high voltage (10 kV) mildly condensed (0.002 μF) repetitive spark discharge in pure helium. Tank

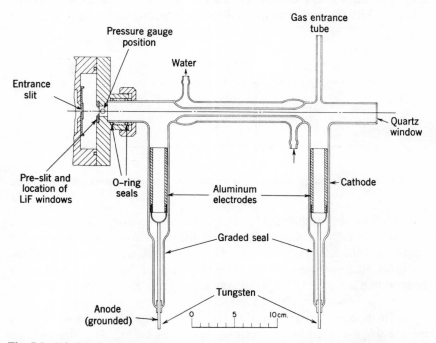

Fig. 5.5 Discharge tube suitable for the production of the rare gas continua (courtesy R. E. Huffman et al. [9]).

helium is purified by passing it through several charcoal or zeolite traps cooled by liquid nitrogen. To remove traces of hydrogen, the helium can be passed through CuO heated to a dull red color, then through a zeolite trap. The purified helium then flows through the windowless light source (the pressure within the light source being approximately 40 torr). The light source used by Tanaka and associates is shown in Fig. 5.5. With the exception of the electrodes and the graded seals, the discharge tube is constructed entirely of quartz tubing. After about every 125 hours of operation, it is necessary to clean deposits from the electrodes with a sodium hydroxide solution.

The excitation circuit is shown schematically in Fig. 5.6. The series spark gap breaks down when the voltage across the capacitor is sufficiently great. The time required to charge the capacitor is determined by the RC time constant of the charging circuit. Thus, the repetition rate of the discharge depends on the value of the product RC. The series spark gap can be replaced by a thyratron circuit producing a more easily controlled discharge [11]. The optimum pulse repetition frequency is about 5 kc per sec. Suitable storage capacitors are the mica type designed for transmitters. With these optimum conditions, the helium continuum, as shown

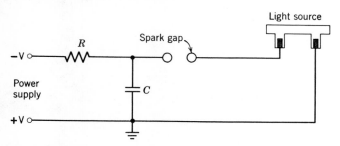

Fig. 5.6 Spark discharge circuit. The spark gap can be replaced by a thyratron (see ref. [11]).

in Fig. 5.7, can be produced [11]. The maximum intensity of the continuum is approximately 10^8 photons per sec at the exit slit of a monochromator operating with a 0.5 Å bandpass. This value will, of course, depend on the individual monochromator. The amplitude of the peak current in the light source is about 100 A, while the average dc current drawn from the power supply is about 100 mA.

Huffman has shown that the Hopfield helium continuum is most intense when excited by a condensed spark discharge. The high voltage condensed ac or dc disruptive discharge with no series spark gap produces a much less intense continuum, while the continuum is practically nonexistent when powered by microwaves of 2450 Mc per sec at power levels up to 800 watt.

That the continuum originates from the helium molecule appears to be well substantiated. Observing time-resolved spectra, Huffman et al. [10] have shown that the continuum light pulse begins approximately 0.14 μsec following termination of the current pulse and reaches a maximum intensity after 1.4 μsec. The light pulse decays slowly with a total effective duration of about 16 μsec. The total duration of the discharge current in these experiments was 0.2 μsec. The observation that the continuum is emitted as an afterglow supports the explanation that the continuum originates from transitions of excited helium molecules. Normally, the emission spectra of molecules are produced more readily in glow discharges than in disruptive spark discharges that tend to produce more ionization and dissociation. However, since the helium continuum is best produced by a spark discharge, this suggests that the atomic helium ion is necessary for the eventual production of the excited helium molecule He_2^*. The

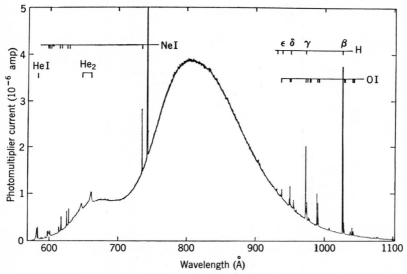

Fig. 5.7 Helium continuum using thyratron modulator. Conditions: sodium salicylate photomultiplier detector (EMI 9514S) at 1630 V; 0.5 Å bandwidth using 100-μ slits; power supply at 116 mA and 10 kV; 5-kc sec pulse repetition frequency; 44-mm Hg helium pressure; 1.0-sec time constant (courtesy R. E. Huffman et al. [11]).

following chain of reactions is suggested as a mechanism operative in a spark discharge in helium.

$$2He + He^+ \rightarrow He_2^+ + He \tag{5.2}$$

$$He_2^+ + e \quad \rightarrow He_2^* \tag{5.3}$$

$$He_2^* \qquad \rightarrow He_2 + h\nu \tag{5.4}$$

$$ \quad \hookrightarrow He + He \tag{5.5}$$

Reaction (5.2) is known to occur readily [15]. The number of excited molecules produced by the above reactions will increase with the total pressure of helium; thus the intensity of the continuum would also be expected to increase with pressure. This is observed to be the case up to a point where competing processes presumably become important. Another fact which suggests that the He^+ ion is the primary reactant is the necessity for extremely pure helium in producing the continuum. In a gas discharge, the energy is absorbed primarily by the species with the lowest ionization potential, and since He has the greatest ionization potential of all the elements (25.58 eV), any impurity will tend to decrease the number of He^+ ions. Introducing about 7 per cent argon impurity into a helium discharge, Huffman et al. [14] completely quenched the continuum.

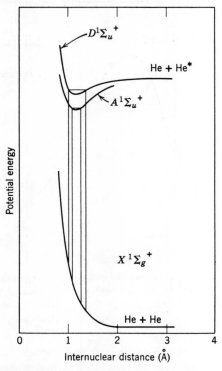

Fig. 5.8 Schematic diagram of the potential energy curves of He_2.

Tanaka et al. [8] have attributed the origin of the two main peaks near 810 and 680 Å to the transitions $A\,^1\Sigma_u^+ \to X\,^1\Sigma_g^+$ and $D\,^1\Sigma_u^+ \to X\,^1\Sigma_g^+$, respectively, as shown schematically in Fig. 5.8. The ground state is repulsive in character, and thus a continuum is emitted by these transitions. Smith and Meriwether [16] have calculated the intensity distribution of the Hopfield continuum using (5.1) and have shown that the transition $A\,^1\Sigma_u^+\,(v = 0) \to X\,^1\Sigma_g^+$ can account for most of the observed continuum.

Another continuum in helium extending from 1050 to 4000 Å has been discovered by Huffman et al. [17,18]. The excitation of this continuum differs from that described above in that the series spark gap must be short circuited, and the helium pressure in the light source must exceed 80 torr.

Under these conditions, when the voltage from the power supply exceeds the breakdown voltage of the light source ($\sim$10 kV), a repetitive discharge is generated even although a dc voltage is applied to the source. The repetition rate is from 100 to 1600 pulse per sec, depending upon the helium pressure and the charging parameters used. The value of R was typically 50 kΩ with C of the order of 0.06 μF. The continuum shown in Fig. 5.9 increased in intensity as the helium pressure was increased up to pressures of 800 torr. This continuum could not be excited when the

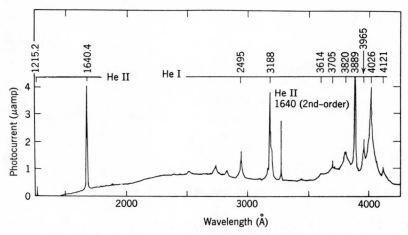

Fig. 5.9 Helium continuum between 1200 and 4000 Å. Pressure 600 torr (courtesy R. E. Huffman et al. [18]).

series spark gap was used. At present, no suitable explanation of the mechanism producing the continuum is available. Similar "new" continua have also been found in argon and xenon.

Several years after Hopfield discovered the helium continuum, Takamine and associates [19] found an emission continuum for neon extending from 744 and 790 Å. They used pressures less than 20 torr. Later, Tanaka et al. [8,20] showed that with higher neon pressures, the continuum extended to longer wavelengths; and at a pressure of 350 torr, the neon continuum covered the range 744 to 1000 Å. Because this falls within the range of the Hopfield continuum, the use of neon has little practical value. In 1954, Tanaka and Zelikoff [21,22] discovered a continuum in xenon and krypton using a microwave discharge. Tanaka [23] extended this work to include argon and to show that the intensity of the continua when excited by a disruptive discharge was roughly twice as intense as that produced by the microwaves. The disruptive discharge, as used in these experiments by Tanaka, was excited with a circuit similar to that shown in Fig. 5.6, but with the exception that no series resistor was used, and that the power supply was a commercial sign transformer rated at 15 kV ac and 900 watt.

Wilkinson [24–26] has described the construction of a simple light source for use with microwave excitation that readily produces continua in Ar, Kr, and Xe. The microwave cavity and light source are shown in Fig. 5.10. The cavity is designed to operate from a 2450 Mc per sec microwave generator rated at 125 W (Burdick Model MW 1200). The appendix tube attached to the light source contains a gettering material such

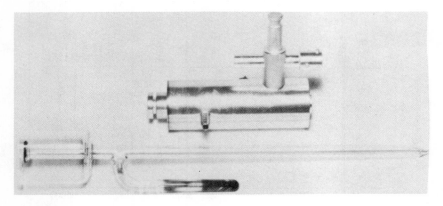

Fig. 5.10 Microwave cavity and light source suitable for the production of the rare gas continua. The appendix tube on the light source contains the gettering material (courtesy GBL Associates).

as metallic barium. The tube is first outgassed under vacuum and the getter is then fired by external heating. Finally, the tube is filled to approximately 200 torr with the rare gas then sealed off. The inclusion of the gettering tube is absolutely essential to reduce the amounts of impurities within the gas. Figure 5.11 shows typical continua in argon, krypton, and xenon. These curves have been corrected for the transmittance of the LiF window and the reflectance of the particular grating used. Wilkinson [26] has shown that the intensity of the continua increases nearly linearly with power using the tuneable cavity. Use of an 800-W generator nearly doubles the intensity of the continuum. Unfortunately, the cavity used with the 800-W generator was not so efficient as the tuneable cavity, so a direct comparison between the 125 W and the 800-W generators was not possible. However, Huffman [14] has shown that even the 800-W microwave generator produces the Ar, Kr, and Xe continua less efficiently than the condensed spark discharge. In a direct comparison with the two methods of excitation he found the argon continuum was about 60 times

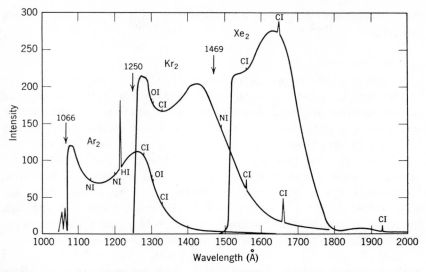

Fig. 5.11 Intensity distribution in the continuous emission spectrum of argon, krypton, and xenon as excited by a 125-W microwave generator. Gas pressure was 200 torr. Atomic emission lines of hydrogen, oxygen, nitrogen, and carbon are indicated (courtesy P. G. Wilkinson).

more intense when excited by the condensed spark discharge. The Hopfield continuum was not observed at all at any pressure using the 800 W microwave generator.

The above continua are believed to be caused by transitions between the excited $^3\Sigma_u^+$ and the unstable $^1\Sigma_g^+$ ground state of the argon, krypton, and xenon molecules. Although these are intercombination transitions, they are less strictly forbidden for the heavier elements.

In summary, the rare gas continua are best produced by a condensed spark discharge using highly purified gases and a bakeable light source. The wavelength range of the continua is given in Table 5.1 along with

Table 5.1   Rare Gas Continua

Gas	Useful Range (Å)	Optimum Pressure Range (torr)	Maximum Intensity[a] (photons/sec)
He	580–1100	40–55	~10^8
Ne	740–1000	>60	
Ar	1050–1550	150–250	~10^8
Kr	1250–1800	>200	~10^7
Xe	1480–2000	>200	~10^7

[a] At the principal maximum with a 0.5 Å bandwidth.

the optimum operating pressures when used with the condensed spark discharge.

Lyman Continuum

Historically, the continuous spectrum emitted by the hydrogen molecule was the first to be used as a practical continuous background source. In fact, it remained the only source of continuous radiation in the vacuum uv region until 1924, when Lyman [27] reported on his flash tube. In this tube, the radiation was produced by discharging a capacitor of about 0.5 μF through a low pressure gas contained in a glass capillary of internal diameter 1 mm. An external spark gap was used to trigger the discharge. Lyman was able to produce a continuum over a wavelength range extending from the visible to about 900 Å. He emphasized that, although the nature of the electrode material and the gas filling were not important, it was essential to use a narrow bore capillary.

Some changes to Lyman's basic design were made by Worley [28] in 1942, but these were concerned mainly with the problem of capillary breakage and the clogging of the spectrograph slits. Worley's description of the Lyman source is as follows:

"When employed as a source of ultraviolet light, the tube usually consists of a thick walled capillary of Pyrex or quartz about 25 mm long and of 1 to 3 mm internal diameter through which is passed intermittently a damped, oscillatory current of large magnitude, derived from a condenser charged to several millicoulombs. Some experience is required to avoid shattering a Pyrex capillary, but when the discharges are properly regulated, it bores out gradually until the walls become too thin for safe use. The capillary must then be replaced."

The problem of the capillary wearing rapidly and finally breaking is the major disadvantage of the Lyman source. Garton [29,30] has overcome this difficulty by violating the premise that the capillary bore must be narrow to produce a continuum. The important parameter for the appearance of a strong continuous emission spectrum is the current density. With current densities around 30,000 A cm^{-2}, the emission is predominantly continuous. At lower current densities, a line spectrum is obtained. Garton attained the necessary high current density by keeping the inductance in the circuit to a minimum. The flashtube had a coaxial construction, and the capacitor was specially designed to have low inductance. The need for low inductance to provide high current densities can be seen from the following equation for an oscillatory capacitor discharge:

$$i = \frac{Vo}{\omega L} \exp\left(-\alpha t\right) \sin \omega t, \tag{5.6}$$

where i is the instantaneous current at a time t; Vo is the capacitor voltage; ω equals $2\pi f$, where f is the ringing frequency of the discharge; L is the total inductance of the circuit; and α is a decay constant equal to $R/2L$, where R is the total resistance of the circuit. Thus for small R

$$i_{max} \approx \frac{Vo}{\omega L}, \qquad (5.7)$$

and since $\omega \approx 1/\sqrt{LC}$ the maximum current increases as L decreases.

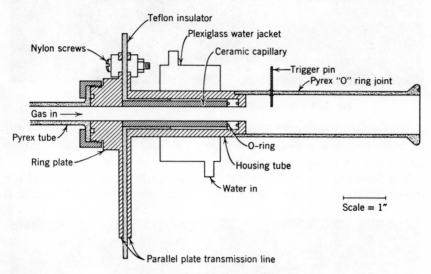

Fig. 5.12   Garton-type flashtube.

Figure 5.12 shows an assembly drawing of the Garton flashtube. The housing tube is the ground electrode, whereas the ring plate is connected to a positive voltage. The pressure inside the capillary is maintained at approximately 20 microns, which is sufficiently low that no spontaneous discharge takes place. The trigger electrode is then used to initiate the discharge. Helium gas purified by flowing through a liquid nitrogen cooled zeolite trap is used since it is the most transparent gas in the vacuum uv region. However, the continuum is produced regardless of the type of gas used, and the line spectrum is due to impurities present in the tube, namely, O_2, Si, and C. The discharge takes place through a ceramic capillary 6.5 cm in length and with a 9 mm bore. The capillary is sealed to the electrode by O-rings. The O-rings limit the operating temperature of the flashtube, which has to be water-cooled if operated as frequently as 1 pps. Ceramic-to-metal seals would be an obvious improvement, although more

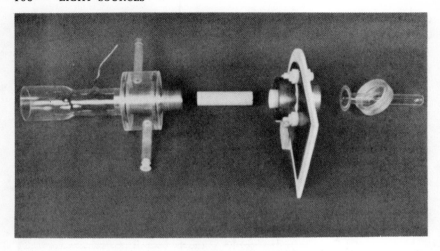

Fig. 5.13 Breakdown of major components of flashtube.

inconvenient to renew. A breakdown of the major components of the flashtube is shown in Fig. 5.13. Figure 5.14 shows an assembled flashtube complete with Plexiglass water jacket.

The flashtube can be automatically or manually fired. In Fig. 5.15, a circuit diagram is given illustrating the trigger circuit and how it is

Fig. 5.14 Assembled view of flashtube.

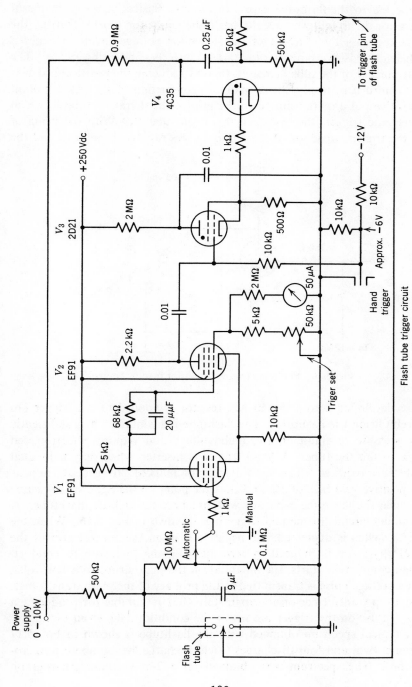

Fig. 5.15 Flashtube trigger circuit for automatic control.

109

connected to the flashtube capacitor circuit. Basically, the principle of operation is as follows. As the flashtube capacitor (9 μF) charges, the voltage across the 0.1 MΩ of the capacitor voltage divider reaches a sufficiently positive potential to start V_1 into its conducting phase. The plate current of the tube increases, thereby reducing the positive grid bias of V_2 until it is insufficient to maintain conduction in V_2. At this point a positive pulse is transmitted to the grid of the thyratron tube V_3 which ignites and triggers the hydrogen thyratron tube V_4. With the firing of V_4, the trigger capacitor (0.25 μF) discharges rapidly through it, and the

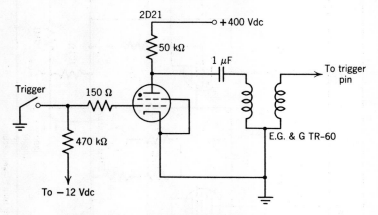

Fig. 5.16   Flashtube trigger circuit for manual control.

pulse developed across the 50 KΩ resistor is applied to the trigger pin thereby firing the flashtube. The flashtube is maintained at a sufficiently low pressure to stand off the high voltage and requires the triggered pulse to fire the tube. A hand trigger is inserted when only individual flashes are required. The variable resistor marked "trigger set" adjusts the positive grid bias of V_2, and thus the point on the *RC* charging curve at which the flashtube capacitor will discharge through the flashtube. A simplified circuit for manual triggering is shown in Fig. 5.16. When the trigger switch is depressed, the negative bias on the control grid of the 2D21 thyratron is reduced to zero allowing the thyratron to conduct. When conduction starts, the potential on the plate drops to a few volts, and a voltage pulse is transmitted to the primary of the pulse transformer. A step-up transformer of approximately 40:1 is suitable for producing a 15-kV pulse on the trigger pin under the conditions shown in Fig. 5.16.

A typical spectrum obtained with the flashtube is shown in Fig. 5.17 for one, two, and four discharges (each discharge lasting about a microsecond). The spectrum was obtained on a 2-m vacuum spectrograph

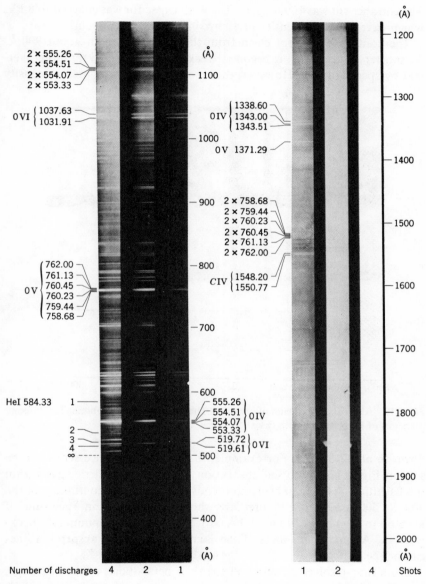

Fig. 5.17 Lyman continuum between 500 and 2000 Å using a Garton-type flashtube with about 20 micron pressure of helium. Exposures for one, two, and four discharges are shown. $C = 9\,\mu\text{F}$; $V = 8\,\text{kV}$; $i_{\max} = 90\,\text{kA}$ per cm².

111

(McPherson No. 240) with a 600-lines-per-mm grating blazed for 1500 Å. The entrance slit was 100 μ wide. The 9-μF capacitor was charged to 8 kV, which gave a current density of approximately 90 kA per cm^2.

The continuum is very clean from the visible down to about 900 Å. At shorter wavelengths, it becomes weaker and progressively more overlaid by spectral lines. However, the continuum is still of useful intensity

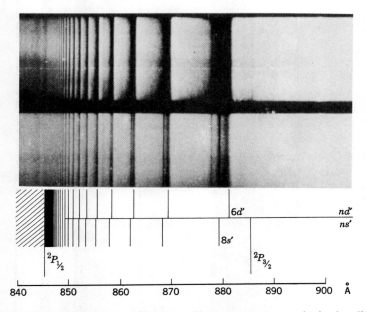

Fig. 5.18 Absorption spectrum of krypton. The two spectra were obtained at different pressures of krypton. In both cases 15 flashes were used.

down to about 300 Å. For example, the helium resonance series can be seen in Fig. 5.17 due to self-absorption by the helium carrier gas within the flashtube. This self-absorption tends to weaken the continuum in the 400 to 500 Å region. Figure 5.18 shows an absorption spectrum of krypton in the region 840 to 890 Å with the Lyman continuum as a background. About 17 members of the Beutler autoionized absorption lines can be seen.

From time-resolved spectra and streak camera studies of the discharge, there is evidence that the discharge current tends to pinch in Garton's coaxial design. This possibly explains the extremely long lifetime of the capillaries [30,31]. The origin of the Lyman continuum has never been determined. Lyman suggested that the continuum owed "its existence to the disintegration products of the glass set free by the erosive action of the discharge." Certainly the carrier gas has no effect on the continuum except at short wavelengths where self-absorption is important. Garton and associates have shown that most of the continuum is emitted from the pinched discharge, while the line spectrum is predominant before the pinch takes place. Although the wall erosion is small in the Garton flashtube, presumably enough enters the main discharge to contribute to the continuum. In fact, recent modifications of the Lyman source allow operation under a high vacuum using no carrier gas [32,33]. In one case, the design shown in Fig. 5.12 can be used by replacing the ceramic capillary with a plastic capillary and by eliminating the trigger pin. When the applied voltage is sufficiently large, a sliding spark is apparently initiated along the inside wall of the plastic capillary. This spark vaporizes some material from the walls, providing a carrier for the main discharge. In another case, a ceramic or teflon capillary is used, and the trigger electrode retained. The trigger electrode is, in this case, designed to form a localized sliding spark to create the necessary ionization for the initiation of the main discharge. These high vacuum flashtubes produce continua similar to that of the Garton flashtube.

The short duration of the light pulse produced by the flashtube ($\sim$1 μsec) and the ability to trigger the discharge at any precisely required time make this type of light source particularly suitable for absorption studies in transient phenomena such as in shock tube research and possibly for the study of reactions greater than 1 μsec.

Synchrotron Radiation

From the classical theory of electricity and magnetism, it is well known that an accelerating electron should radiate energy. Visible radiation was actually observed from accelerating electrons soon after the construction of high-energy electron accelerators. In fact, as higher energy machines were constructed, it became clear that the radiative losses incurred by the electron placed a severe limitation on its final energy. This drawback to high-energy physics is advantageous to spectroscopy, since the radiation emitted is continuous from the visible into the x-ray region. Thus, in principle, it provides the ideal light source.

Much theoretical work has been performed on the radiation from centripetally accelerated electrons [34–41]. Experimentally, the first investigation was conducted in 1947 on the 70 MeV General Electric Synchrotron [42,43]. This investigation was in the visible portion of the spectrum. The results agreed with the spectral distribution formulae based on the extensive classical treatment of the problem by Schwinger [41]. The first use of synchrotron radiation in the soft x-ray region was reported by Hartman and Tomboulian in 1953 [44]. Subsequent reports by Tomboulian and co-workers have covered both the theoretical and experimental aspects [45,46]. The use of the 180 MeV synchrotron at the National Bureau of Standards as a spectroscopic light source for radiation between 100 and 600 Å has been described by Codling and Madden [47]. Cauchois et al. [48] have utilized a 1.1 GeV synchrotron to study the K-absorption and to excite the K-radiation of aluminum in the 5 to 14 Å band.

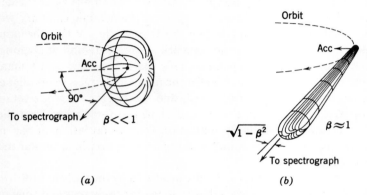

Fig. 5.19 Qualitative radiation patterns to be expected from electrons in a circular orbit (*a*) at low energy and (*b*) as distorted by relativistic transformation at high energy. See also discussion in W. K. H. Panofsky and M. Phillips, "Classical Electricity and Magnetism." Cambridge: Addison-Wesley, 1962, 2nd ed., pp. 363–370 (courtesy D. H. Tomboulian and P. L. Hartman [45]).

Qualitative radiation patterns to be expected from electrons in a circular orbit are shown in Fig. 5.19 [45,49]. As the energy of the electron increases, the emitted radiation becomes more directional, until at relativistic energies, the radiation is confined to a narrow cone pointing in the direction of the electron's motion and confined to the orbital plane. The angular distribution of the radiation is a function of both the electron energy and the spectral region. At a fixed energy, the angular distribution increases with wavelength, as shown in Fig. 5.20 for a monoenergetic electron of 180 MeV. However, the distribution in which the major portion of the radiation is contained is given, approximately, by $2m_0c^2/E$, where E is the electron energy and m_0c^2 its rest energy. When E is measured in MeV, the angular distribution is just $1/E$. For a 180 MeV synchrotron, the radiation is, therefore, confined to a cone with a divergence of about 5.5×10^{-3} radians. Thus the radiation falling on a diffraction grating has very little vertical divergence and is nearly parallel to the Rowland plane. Under these conditions, there is very little astigmatism present in grazing incidence spectrographs as shown by Tomboulian and Hartman [45].

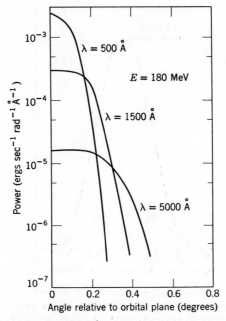

Fig. 5.20 Power (erg sec^{-1} rad^{-1} Å^{-1}) radiated per electron as a function of the observation angle ψ measured relative to the orbital plane, for monoenergetic electrons of 180 MeV energy and three wavelengths (courtesy K. Codling and R. P. Madden [47]).

A potentially useful feature of the synchrotron radiation is the fact that it is plane polarized when viewed in the orbital plane, the electric vector being parallel to this plane. However, when the cone of radiation is viewed at an angle ψ to the orbital plane, the light becomes elliptically polarized and eventually circularly polarized as ψ becomes large [47,50]. Figure 5.21 shows the intensity cross section of the radiation cone for radiation polarized in the orbital plane ($I_{\parallel}$) and perpendicular to it ($I_{\perp}$). The solid lines represent theory, while the circles are the experimental points.

The instantaneous power radiated into all angles per unit wavelength interval by a single accelerating electron is given by [45]

$$P(\lambda, t) = \frac{3^{5/2}}{16\pi^2}\left(\frac{e^2 c}{R^3}\right)\left[\frac{E(t)}{m_0 c^2}\right]^7 G(y), \qquad (5.8)$$

where $E(t)$ is the instantaneous energy of the electron, $m_0 c^2$ is the rest energy of the electron, and R is the radius of the electron orbit. The function $G(y)$ is defined as

$$G(y) = y^3 \int_y^\infty K_{5/3}(\eta)\, d\eta, \qquad (5.9)$$

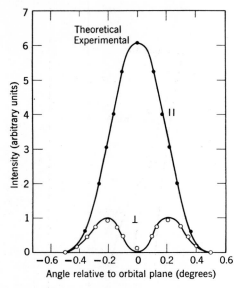

Fig. 5.21 Comparison of theory and experiment for the intensity radiated in each component of polarization as a function of observation angle ψ measured relative to the orbital plane, for monoenergetic electrons (electron energy = 120 MeV; $\lambda = 5000$ Å). The experimental data are normalized to the theoretical curves at one point only, that is, the peak of the parallel component distribution ($\parallel$–component parallel to the orbital plane, $\perp$–component perpendicular to the orbital plane). (Courtesy K. Codling and R. P. Madden [47].)

where

$$y = \frac{\lambda_c}{\lambda} \quad \text{and} \quad \lambda_c = \frac{4\pi R}{3}\left[\frac{m_0 c^2}{E(t)}\right]^3. \quad (5.10)$$

The integrand in (5.9) involves Bessel functions of imaginary argument and fractional order as follows:

$$K_{5/3}(\eta) = \frac{\pi}{\sqrt{3}}\frac{4}{3\eta}[i^{2/3}J_{-2/3}(i\eta) - i^{-2/3}J_{2/3}(i\eta)]$$
$$+ [i^{1/3}J_{-1/3}(i\eta) - i^{-1/3}J_{1/3}(i\eta)]. \quad (5.11)$$

The integral can be evaluated numerically when the values of the Bessel functions [51] are substituted in (5.11). The value of $G(y)$ is plotted as a function of $(1/y)$ in Fig. 5.22. Using (5.10), the value of $G(y)$ can be obtained from Fig. 5.22 for any wavelength; substituting this value of $G(y)$ into (5.8) gives the value of $P(t)$ at that wavelength. Figure (5.23) illustrates the relative spectral distributions of the power radiated by monoenergetic electrons. The wavelength corresponding to the peak of the

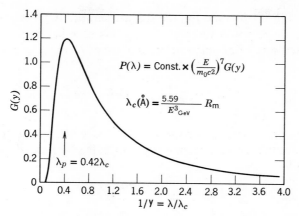

Fig. 5.22 Universal spectral distribution curve for the radiation from monoenergetic electrons. $P(\lambda)$ is the instantaneous power radiated per angstrom and λ_c, expressed in angstrom units, is given by (5.10). (Courtesy D. H. Tomboulian and P. L. Hartman [45].)

distribution is given by $\lambda_p = 0.42\lambda_c$; that is,

$$\lambda_p \ (\text{Å}) = \frac{23.48 \times 10^8}{E^3(\text{MeV})} R(m), \tag{5.12}$$

where λ_p is measured in angstroms, $R(m)$ in meters, and $E(\text{MeV})$ in units of MeV.

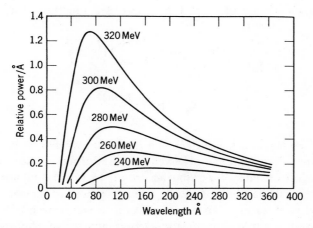

Fig. 5.23 Comparison of the relative spectral distributions for the instantaneous power radiated by monoenergetic electrons at various energies. This illustrates the sensitive variation of the distribution with energy (courtesy D. H. Tomboulian and P. L. Hartman [45]).

If the radiation is observed for the full acceleration interval, we must consider the average power radiated during this time in order to find the spectral distribution. For an acceleration time T, the average power $\bar{P}(\lambda)$ is given by

$$\bar{P}(\lambda) = \frac{1}{T} \int_0^T P(\lambda, t) \, dt. \tag{5.13}$$

Consider a specific accelerator such as the Cambridge Electron Accelerator, where the instantaneous electron energy $E(t)$ has a time dependence represented by

$$E(t) = E_0 \sin^2 \left(\frac{\pi t}{2T} \right), \tag{5.14}$$

where E_0 is the maximum energy of the electron, and T is equal to one quarter of the periodic cycle and represents the full acceleration interval. Tomboulian and Bedo [46] have evaluated (5.13) using (5.8). They obtained for a single electron

$$\bar{P}(\lambda) = \frac{3^{3/2}}{8\pi^3} \left(\frac{e^2 c}{R^3} \right) \left(\frac{E_0}{m_0 c^2} \right)^7 x^4 L(x), \tag{5.15}$$

$$= 1.6 \times 10^{-22} \frac{E_0^7 (\text{MeV})}{R^3 (m)} \, x^4 L(x) \text{ erg sec}^{-1} \, \text{Å}^{-1}. \tag{5.16}$$

The function $x^4 L(x)$ is plotted in Fig. (5.24) as a function of x, where

$$x = \frac{\lambda_0}{\lambda} \quad \text{and} \quad \lambda_0 = \left(\frac{4\pi R}{3} \right) \left(\frac{m_0 c^2}{E_0} \right)^3, \tag{5.17}$$

or

$$\lambda_0 \, (\text{Å}) = 55.9 \times 10^8 \frac{R(m)}{E_0^3 (\text{MeV})}. \tag{5.18}$$

The function $x^4 L(x)$, and hence $\bar{P}(\lambda)$, is a maximum when $\lambda = \frac{1}{2}\lambda_0$. This is true only when the radiation is observed over the full acceleration interval and for an electron energy varying according to (5.14). Figure 5.25a shows the spectral distribution of the average power $\bar{P}(\lambda)$, when E_0 is 6 BeV and R is 26.26 m. The power peaks at a wavelength of 0.34 Å. For comparison, a curve showing the average number of photons $\bar{N}(\lambda)$ emitted per second in a 1 Å band is also included in the figure. Figure 5.25b shows the continuation of this curve to longer wavelengths. For

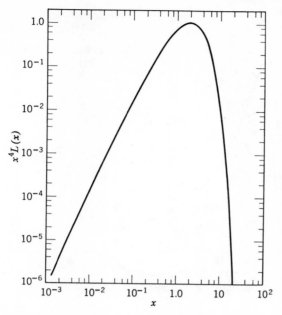

Fig. 5.24 A universal curve representing the spectral distribution on a logarithmic scale. For specified values of the peak energy and orbital radius, the quantity $x^4L(x)$, plotted as ordinate, is proportional to the average radiated power per angstrom. The dimensionless variable x ($x = \lambda_0/\lambda$) is inversely proportional to the emitted wavelength λ. The power is averaged over the full acceleration interval during which the electron energy increases according to $E_0 \sin^2(\pi t/2T)$. (Courtesy D. H. Tomboulian and D. E. Bedo [46].)

an electron beam containing 10^{10} electrons, the radiated flux at 200 Å would be about 3×10^{15} photons-sec^{-1} Å^{-1}.

For wavelengths shorter than 600 Å, the synchrotron is one of the most useful sources of continuous radiation. Thus Madden and Codling [52–56] were able to obtain the first absorption spectra of the rare gases below 600 Å. Much new structure was observed. Figure 5.26 is a reproduction of their results for the absorption of He, Ne, and Ar in the wavelength range 180 to 425 Å [52].

Bremsstrahlung Radiation

When high-speed electrons are stopped abruptly by a target, there are, in general, two types of spectra emitted in the x-ray region, a continuous spectrum and a line spectrum. The line spectrum, or characteristic radiation, is dependent on the nature of the target, since the emission is caused

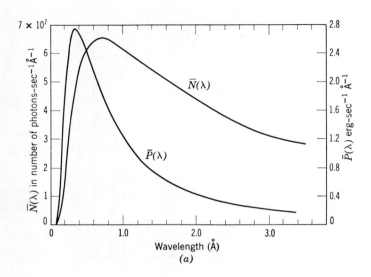

(a)

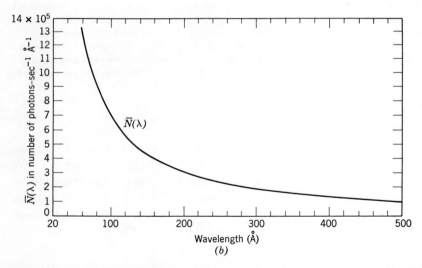

(b)

Fig. 5.25 *(a)*, *(b)* The average power $\bar{P}(\lambda)$ and the average number of photons $\bar{N}(\lambda)$ emitted per second per angstrom band over a full acceleration interval for $E_0 = 6$ BeV and $R = 26.26$ m. $\bar{P}(\lambda)$ is obtained from Fig. 5.24 by multiplying $x^4 L(x)$ by 2.46 (courtesy D. H. Tomboulian).

120

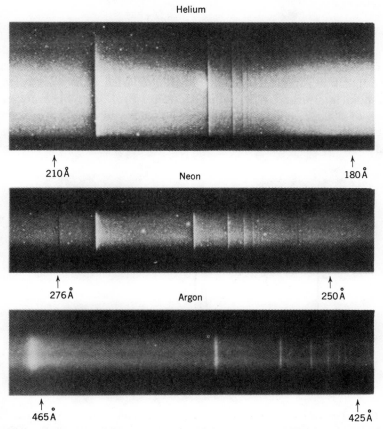

Fig. 5.26 Absorption spectra showing discrete resonance structure in the photo-ionization continuum of He, Ne, and Ar (courtesy R. P. Madden and K. Codling [52]).

by electronic transitions within the atoms composing the target. Although the potential difference accelerating the electrons towards the target must be at least a certain value in order to excite these characteristic lines, the wavelengths are entirely independent of this potential difference. The continuous spectrum, on the other hand, is dependent on the potential difference. Figure 5.27 illustrates the typical form of an x-ray spectrum showing the characteristic radiation superimposed on the continuum. As the accelerating voltage is decreased, the abrupt short wavelength cut-off of the continuum moves towards larger wavelengths. The short wavelength limit λ_{min} is determined from the relation

$$eV = h\nu_0. \tag{5.19}$$

Therefore

$$\lambda_{min}(\text{Å}) = \frac{12400}{V(\text{volt})}, \qquad (5.20)$$

where V is the potential difference in volts across the x-ray tube and ν_0 is the maximum frequency of the radiation emitted. The wavelength of maximum intensity is given (approximately) by $1.5\lambda_{min}$ for voltages in excess of 2000 V. At lower voltages, however, the maximum tends to shift towards λ_{min} approaching the spectral distribution expected from an ideal thin target in which the intensity rises immediately to a maximum

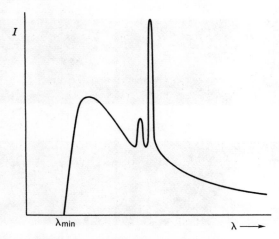

Fig. 5.27 Typical spectrum of bremsstrahlung radiation showing the characteristic line radiation superimposed on the continuum. λ_{min} indicates the short wavelength limit of the spectrum.

at λ_{min} and falls off to longer wavelengths as $1/\lambda^2$ [57,58]. Figure 5.28 shows the energy distribution in the continuous x-ray spectrum between 6 and 22 Å, as obtained by Neff using a platinum target and a variety of accelerating potentials [59]. The voltages were too low to produce characteristic radiation. Such continua, as shown in Fig. 5.28, do not have a long wavelength cut-off and, consequently, can be useful at much longer wavelengths. Peterson and Tomboulian [60,61] have made use of the long wavelength tails in studying the x-ray continuum between 80 and 180 Å and have found that the decrease in intensity I with wavelength follows the relation

$$I = \frac{C}{\lambda^a}, \qquad (5.21)$$

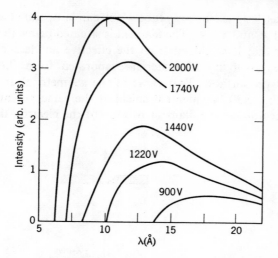

Fig. 5.28 Energy distribution in the continuous x-ray spectrum for a variety of accelerating potentials. A platinum target was used (courtesy H. Neff [59]).

where C is a constant, and the value of a varies from about 1.8 for elements of low atomic number to 3.0 for high atomic numbers. They also found that the value of a varied with the accelerating voltage. Lukirskii and Fomichev [62] have also made use of the x-ray continuum from 25 to 250 Å. Both groups operated their x-ray tubes typically around 3 kV and 200 mA using water-cooled targets. The intensity of the x-ray continuum increases linearly with the electron current. This would suggest that low voltage and high current x-ray tubes could provide excellent continua. However, the maximum current which can be used at a given voltage is space charge limited. Techniques to circumvent this problem may be devised, but at present, little work has been done in this direction.

Designs for x-ray sources have been discussed by several authors [61,63–65]. A schematic diagram of the principal components of an x-ray tube is shown in Fig. 5.29. The x-ray emission is frequently observed at an angle θ of about 90° measured from the direction of the incident electron beam. If the incident electrons were all decelerated along their direction of incidence, the angular distribution of the resulting radiation would have an intensity maximum at 90° according to classical electromagnetic theory. However, Stephenson [57] has reported a maximum intensity at 50° for 38 kV electrons and at 65° for 16 kV electrons. Quantum mechanics predicts an intensity maximum at about 55°. The above angles are all measured from the direction of the incident electrons.

Henke [63] has designed an x-ray tube which can be operated with anode currents up to 1 A at 3000 V. The filament is situated below the anode in such a manner that it is hidden from the effective anode surface, thus eliminating the possibility of material evaporated from the filament reaching the target surface. Two views of the geometrical arrangement are shown in Fig. 5.30. A filament shield at the same potential as the filament is placed above the filament to prevent the electron beam from

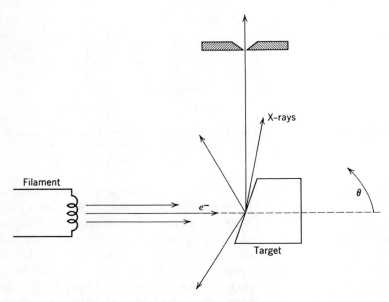

Fig. 5.29   Schematic diagram of the principal components of an x-ray tube.

irradiating the bottom of the anode structure. To prevent a space charge building up around the filament, grids are brought down from the anode. The accelerated electrons pass through the grids and are focused onto the anode by a cylindrical electrode which is at the same potential as the filament. In addition to preventing anode contamination from the filament, this geometrical arrangement allows a large anode current to flow even at low voltages. The anode is cooled to a working temperature of between 300°C and the melting point of the anode target material. This prevents contamination of the anode from impurities within the vacuum system.

The intensity of bremsstrahlung radiation, as produced in conventional x-ray tubes, is sufficiently weak that photographic recording or counting techniques are necessary to observe the radiation.

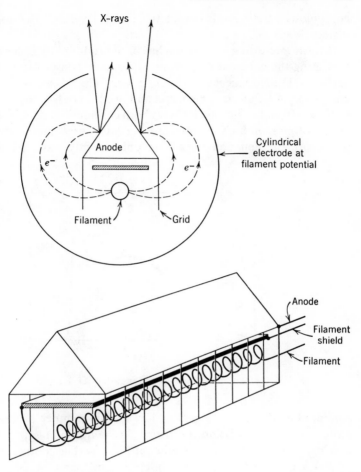

Fig. 5.30 High-current x-ray tube as used by Henke [63]. The filament is concealed from the anode to prevent anode contamination.

Miscellaneous Continua

(a) In connection with our discussion, certain characteristic x-rays form bands of continuous radiation several angstroms in extent [66]. One such band is the Al L_3 emission band shown in Fig. 5.31. This spectrum covers an interval of about 40 Å starting with a short wavelength limit at 170 Å and continuing to about 210 Å. The Mg L_3 band covers the interval 250 to 310 Å. These bands are characteristic of the atoms forming the anode material and originate from electronic transitions between the continuous energy levels within the Fermi level and the appropriate L, M, etc., shells of the atom.

(b) An interesting continuum has been produced by Ehler and Weissler [67] using the coherent light from a Q-switched laser focused on a metal surface. The 100-megawatt peak power ruby laser pulse with a total energy of 0.4 J vaporizes the metal and produces a hot dense plasma. The radiation from this plasma includes both continuous and discrete emission. When targets of high atomic number were used (such as tungsten or platinum), the emission lines were only slightly more intense than the

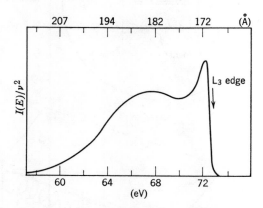

Fig. 5.31 AlL_3 x-ray emission band (courtesy D. H. Tomboulian [66]).

continuum. However, targets of low atomic number produced a much more intense line emission spectrum. The continuum apparently has two maxima, one at 200 Å and the other at 1700 Å. The continuum maximum at 200 Å appeared to be more intense than that at 1700 Å. However, since the photographic plates were not calibrated nor were the reflectances of the gratings known, an accurate comparison between the two maxima was not possible. For free-bound emission, a continuum maximum is expected to occur at a wavelength of $6200/T$ (eV) angstrom units. [68] Thus the plasma produced by the laser pulse would be expected to have an electron temperature of 31 eV based on a continuum maximum of 200 Å.

(c) Visible radiation from an electron passing close to the surface of a metal diffraction grating and moving at right angles to the rulings has been observed by Smith and Purcell [69]. The radiation was observed to be strongly polarized with the electric vector lying in the plane perpendicular to the grating surface. The motion of the electron image is considered to be modulated by the regular undulations of the grating surface, and as the electron's image moves to and from the electron, an oscillating electric dipole is created (see Fig. 5.32). Smith and Purcell have shown that the wavelength of this radiation is given by

$$\lambda = d\,(\beta^{-1} - \cos\theta), \qquad (5.22)$$

where d is the groove separation of the grating; β is equal to v/c, where v is the electron velocity and c is the velocity of light; and θ is the angle

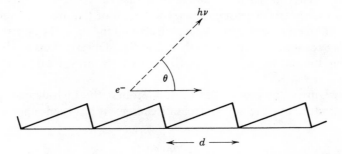

Fig. 5.32 Radiation emitted from a high-energy electron passing close to a metal diffraction grating with grooves separated by a distance d.

between the direction of motion of the electron and the light rays. Thus with an electron beam of fixed energy, the wavelength of the radiation continuously increases as the angle of observation θ increases. With a 340 kV electron beam, Smith and Purcell observed radiation of 4600 Å at an angle of 20° from the grating surface. By changing the energy of the electrons, a continuous spectrum can be scanned passed a fixed observation angle. For use in the vacuum uv spectral region, it is expected that electrons of energies in excess of 1 MeV would be required. The possibility of generating radiation in this manner was apparently first pointed out by Frank [70] and has subsequently been discussed by others [71–74].

(d) While studying radiation from condensed sparks in a vacuum, Balloffet [75] observed a continuum which was present only when electrode materials of high atomic number were used, such as tungsten and platinum. Pursuing this work, Balloffet et al. [76] used a uranium anode, removed the capillary, and observed the radiation emitted in the vicinity of the anode. The continuum extended from 2000 to 80 Å. The short wavelength limit of observation was imposed by the particular grazing incidence spectrograph used and not necessarily by the light source. The discharge conditions were typically 0.5 μF, 22 kV, 0.08 μH, and a peak current of 55 kA.

The pulse duration was 1.25 μsec. Further study by Balloffet et al. [77] has revealed that soft x-rays are also present. Using a crystal of NaCl, they observed Laue diffraction patterns caused by x-rays in the vicinity of 1 Å. The short wavelength limit of bremsstrahlung radiation produced by 22 kV electrons is 0.56 Å, while the wavelength at maximum intensity is approximately 0.84 Å (see Bremsstrahlung Radiation). Thus, it might be expected that these short wavelength x-rays ($\sim$1 Å) are produced by a conventional bremsstrahlung effect.

The spatial distribution of the continuum has been studied by Lotte et al. [78]. They have shown conclusively that the continuum is produced in a plasma sheath extending down the length of the uranium anode for several millimeters, including the region around the tip of the electrode and extending to about 1 mm away from the tip. Observations at distances exceeding 1 mm from the electrode tip revealed the presence of a strong line spectrum.

The origin of the continuum centers around the high-density plasma sheath produced by the high-current discharge. It has been suggested that

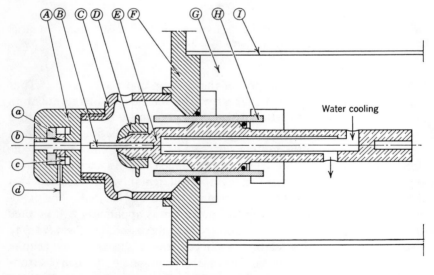

Fig. 5.33 Spark light source with uranium anode for the production of continuous radiation. *A*. Cathode—triggering device assembly: (*a*) insulating washer; (*b*) auxiliary anode; (*c*) thin insulating washer (sliding spark support); (*d*) auxiliary anode lead-in. *B*. Principal anode (uranium). *C*. External conductor; holes are provided for evacuation. *D*. Mandrel. *E*. Water-cooled anode holder. *F*. Flange for the source fixation. *G*. Insulating piece for the centering of the anode-holder. *H*. Rectified alumina tube. *I*. External conductor outside the vacuum. (Courtesy H. Damany, J-Y Roncin, and N. Damany-Astoin [80].)

the deceleration of electrons within this sheath produces bremsstrahlung radiation [76,78,79]. Thus with the extremely high current densities present in spark discharges, a strong continuum might be expected. Contributions from free-bound transitions must certainly be present but to what extent is unknown. The observed facts, however, show that an anode of high atomic number must be used. When aluminum or magnesium electrodes were used, there was no sign of any continuum.

A recent version of this light source has been described by Damany et al. [80], and is shown in Fig. 5.33. The coaxial design is essential to reduce inductance and produce the necessary high-current densities. The water-cooled anode is fitted with a mandrel to hold a tungsten or uranium rod 4 mm in diameter and is insulated from the cylindrical return circuit by a ceramic tube. The high-vacuum discharge is triggered by a sliding spark situated in the cathode. Typical parameters used are $C = 0.05\ \mu F$, $V = 22$ kV, and a vacuum of 10^{-5} to 10^{-6} torr. The source can be fired at a maximum rate of five to ten pulses per sec. The hole through the cathode acts as a diaphragm, the size of which can be adjusted to screen off the undesired regions of the discharge where lines are emitted.

5.3 LINE SOURCES

Line radiation is produced by electronic transitions between different energy levels in neutral atoms and molecules and in ions. The Roman numerals placed after the symbols for the elements indicate whether the radiation is emitted from the neutral atom or from ions of various degrees of ionization. For example, radiation emitted from the neutral oxygen atom is designated O I, while radiation from singly ionized oxygen (one electron removed) is designated O II. Higher degrees of ionization are designated O III, O IV, etc. In general, the shortest wavelengths are produced from the most highly ionized atoms (see Chapter 10). The width or sharpness of the lines depends upon the natural width of the energy levels involved in the transitions and upon external conditions such as pressure, temperature, and electric fields. Typically, line widths range between 0.1 and 0.001 Å. An energy level diagram for the helium ion is shown in Fig. 5.34, illustrating the radiation produced by various transitions. Radiation produced by transitions between excited states and the ground state of the atom or ion is called resonance radiation. The first resonance line (namely, the one produced by a transition from the first excited state and the ground state of the atom or ion) is known as the *raie ultime*. If fine structure exists within the energy level of the first excited state thus producing two or more lines, the most intense of these lines is designated the *raie ultime*. In the case of He II, the wavelength of

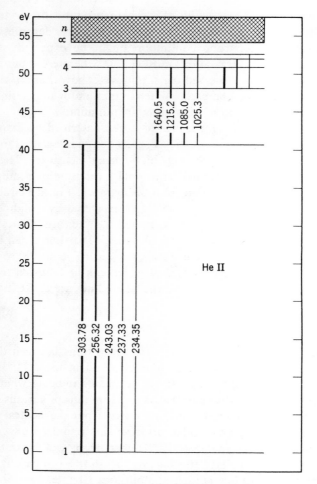

Fig. 5.34 Energy level diagram for He II.

the raie ultime is 303.781 Å. Table 5.2 lists the wavelengths of the first resonance lines emitted from the neutral atoms of the rare gases.

The problem of producing radiation of a given wavelength becomes the problem of creating the appropriate excited atom or ion. Various discharge modes are necessary to produce the different excited species. A dc glow discharge tends to produce radiation from neutral atoms (often referred to as the arc spectrum). When molecular gases are used in a discharge, a molecular band spectrum is often produced. More energetic discharges such as spark discharges are necessary to produce highly ionized atoms. This type of radiation is commonly referred to as the spark

Table 5.2   The First Resonance Lines and Ionization
Potentials of the Rare Gases[a]

Gas	First Resonance Line (Å)	(eV)	Ionization Potential (Å)	(eV)
He	584.334*	21.22	504.259[b]	24.59
Ne	735.895*	16.85	574.938[c]	21.56
	743.718	16.67		
Ar	1048.219*	11.83	786.721[c]	15.76
	1066.659	11.62		
Kr	1164.867	10.64	885.620[c]	14.00
	1235.838*	10.03		
Xe	1295.586	9.57	1022.140[c]	12.13
	1469.610*	8.44		

* Wavelengths of the *raie ultime.*
[a] See Table 10.1.
[b] G. Herzberg, *Proc. Roy. Soc.* (London) **248,** 309 (1958).
[c] C. E. Moore, *Atomic Energy Levels*, Circ. **467,** Vols. I–III, National Bureau of Standards (U.S. G.P.O., Washington, D.C., 1949, 1952 and 1958).

spectrum. The degree of ionization is dependent not only on the voltage used, but on the capacitance and inductance of the discharge circuit.

Many different light source designs are necessary not only for the production of given wavelengths, but also for specialized requirements such as high spectral purity, high intensity, and the production of sharp lines. Some of the most important and commonly used light sources are described below. Pages 131 to 162 describe the various modes for exciting a gaseous discharge, while pp. 163 to 174 describe the spark discharge both in a gas and at high vacuum.

Cold Cathode Discharge

When the voltage across a discharge tube containing a gas at low pressure (∼1 torr) is increased, a point is reached when the discharge "strikes." The striking potential is much greater than that necessary to maintain the discharge. The electrical circuit is shown in Fig. 5.35. Typical parameters are: ballast resistor R ≈ 1000 Ω, striking potential ≈ 2000 V, gas pressure ≈ 1 torr, operating potential of discharge tube ≈ 600 V, and discharge current ≈ 100 to 500 mA. Actual operating conditions depend, of course, on the dimensions of the discharge tube, the type of gas, and the gas pressure.

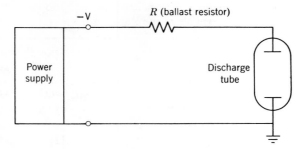

Fig. 5.35   Electrical circuit for a glow discharge tube.

In general, a glow discharge consists of alternate light and dark regions. This is shown schematically in Fig. 5.36. Starting at the cathode is a short region called the Aston dark space. Next to this is a glow called the *cathode glow*, which is followed by another dark region called the *cathode dark space* or the Crookes dark space. This is followed by the *negative glow*, which is long compared with the cathode glow and is most intense on the cathode side. The Faraday dark space follows the negative glow and precedes the *positive column*, which fills most of the remaining length of the discharge. At appropriate pressures, striations appear within the positive column. Towards the anode there may or may not be a bright glow called the *anode glow*. Finally, the *anode dark space* precedes the anode. The presence of all of these features depends critically on the nature of the gas, the pressure, and the magnitude of the discharge current. However, although the positive column and the features towards the anode may disappear under certain conditions, the features at the cathode tend to be permanent since they are involved with the maintenance of the discharge.

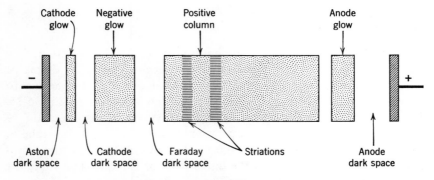

Fig. 5.36   Emission characteristics of a glow discharge.

In the normal glow discharge the potential drop across the discharge (or more precisely between the cathode and the negative glow called the cathode drop) remains constant as the discharge current increases. As the current continues to increase, a value is reached at which point the voltage starts to rise. This region is known as the abnormal glow discharge. A point is finally reached when the voltage has a maximum value and starts to decrease with increasing current until the discharge is being maintained at a very low voltage and high current. This is the region of the arc discharge. The transition from a glow to an arc discharge is usually a rapid and discontinuous process. Compton [81] has defined the arc as "a discharge of electricity, between electrodes in a gas or vapor, which has a voltage drop at the cathode of the order of the minimum ionizing or minimum exciting potential of the gas or vapor." In the glow discharge, most of the voltage appears between the cathode and the start of the negative glow. Thus a very high electric field is set up which produces electrons of sufficient energy to excite not only the neutral atoms but also to excite ions. This excitation process takes place within the negative glow and is responsible for the type of radiation emitted from this region. Although the electrons are again accelerated across the Faraday dark space, the potential drop is much less, and they have insufficient energy to excite ionic species within the positive column. Thus in general, the radiation emitted from the positive column is due to excited atoms (arc spectrum), while the radiation emitted from the negative glow contains contributions from excited ions (spark spectrum) as well as from excited atoms.

The degree of excitation in the negative glow should depend upon the magnitude of the cathode drop. The normal cathode drop depends on the combination of gas and electrode material involved. For example, the normal cathode drop in an argon discharge with aluminum electrodes is about 100 V, but with iron electrodes, it is about 165 V. Table 5.3 lists the approximate values of the normal cathode drop for different combinations of electrode and gas [82]. In general, the cathode drop increases nearly linearly with the work function of the cathode material. As previously stated, the cathode drop further increases with the discharge current when the discharge is operated in the abnormal glow region. In the normal glow discharge, only a portion of the cathode surface acts as an emitter. As the discharge current increases, the current density at the cathode remains constant; however, a larger portion of the cathode surface is now involved, but the cathode drop remains constant. The transition between the normal and abnormal glow regions occurs when the discharge current is increased to such a value that the cathode is completely covered with the negative glow and its entire surface acts as an emitter. The cathode current density then increases, and the voltage drop of the

Table 5.3 Normal Cathode Fall[a]
(volts)

Cathode	Air	Ar	He	H_2	Hg	Ne	N_2	O_2	CO	CO_2	Cl
Al	229	100	140	170	245	120	180	311			
Ag	280	130	162	216	318	150	233				
Au	285	130	165	247		158	233				
Ba		93	86				157				
Bi	272	136	137	240			210				
C				240	475				525		
Ca		93	86			86	157				
Cd	266	119	167	200		160	213				
Co	380										
Cu	370	130	177	214	447	220	208		484	460	
Fe	269	165	150	250	298	150	215	290			
Hg			142		340		226				
Ir	380										
K	180	64	59	94		68	170		484	460	
Mo					353	115					
Mg	224	119	125	153		94	188	310			
Na	200		80	185		75	178				
Ni	226	131	158	211	275	140	197				
Pb	207	124	177	232		172	210				
Pd	421										
Pt	277	131	165	276	340	152	216	364	490	475	275
Sb	269	136		252			225				
Sn	266	124		226			216				
Sr		93	86				157				
Th						125					
W					305	125					
Zn	277	119	143	184			216	354	480	410	
CsO-Cs						37					

[a] A. V. Engel and M. Steenbeck, *Elektrische Gasentladungen, ihre Physik u. Technik*, Vol. 2, p. 103. J. J. Thomson and G. P. Thomson, *Conduction of Electricity through Gases*, Vol. 2, pp. 331–332. Extracted from J. D. Cobine, *Gaseous Conductors* (Dover, New York, 1958).

cathode region also increases as the discharge current increases. Thus if it is desirable to operate the normal glow with large discharge currents, the cathode surface must be made as large as possible. On the other hand, if it is desirable to produce a large cathode drop, the abnormal glow must be used. To reach this condition at low values of the discharge current, a cathode of small area should be used. Normally, however, the light intensity increases with the discharge current. Thus a large cathode surface area is again desirable to provide a maximum discharge current without causing a transition to the arc discharge. Another reason for using cathodes of large surface area is that they erode less rapidly under the continual bombardment of the positive ions. The stability of the radiation emitted from a glow discharge can depend on the smoothness of the cathode surface.

Positive Column.   To increase the intensity of the radiation from a glow discharge, it is customary to confine the discharge in a capillary 10 to 20

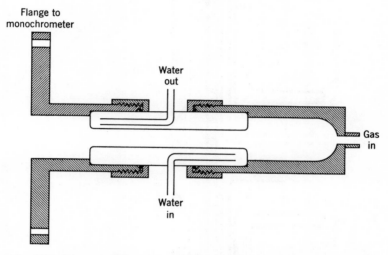

Fig. 5.37   Water-cooled capillary discharge tube with external electrodes.

cm in length with a bore of 3 to 5 mm and to view the discharge "end-on," that is, along the axis of the capillary. The positive column frequently fills the entire length of the capillary and is the main source of the radiation observed. Normally, the voltage gradient along the positive column is only a few volts per cm. However, as the diameter of the capillary decreases, the voltage gradient increases.

A common design for a cold cathode discharge tube is to use a water-cooled quartz or Pyrex capillary sealed into a hollow cathode and anode by O-rings, such as those shown in Fig. 5.37. The electrodes can be air cooled, but it is more efficient to use water cooling. Another version of the water-cooled capillary is shown in Fig. 5.38. The main disadvantage of the water-cooled capillary is the danger of water entering the vacuum spectrograph in the event the capillary breaks. This danger can be circumvented by constructing the discharge tube from quartz, ceramic or

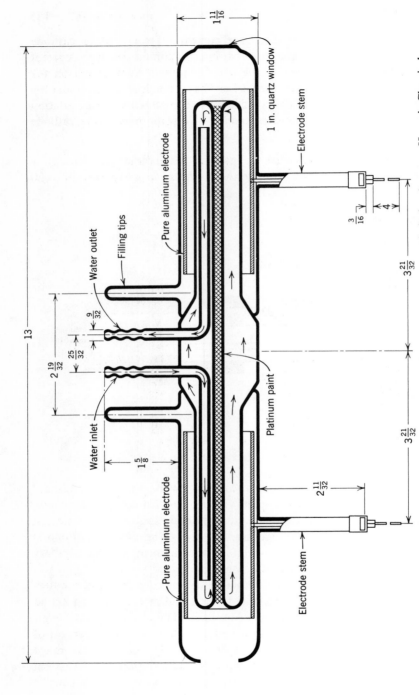

Fig. 5.38 Water-cooled capillary discharge tube with internal electrodes. All measurements are in inches (courtesy Hanovia Chemical Company, N.J.).

136

other high-temperature materials and simply cooling the electrodes. Figure 5.39 shows a breakdown of a discharge tube designed by Hunter [83] in which the quartz capillary is allowed to run hot. Basically, the light source consists of a 4 in. quartz disk with a quartz capillary (4 mm bore) sealed into the disk through its center and normal to its surface. The quartz disk is ground flat to seat the O-rings on the cathode and the anode, and it acts as an insulator as well as a vacuum seal. The cathode (the left-hand cylinder in Fig. 5.39) is constructed of two concentric

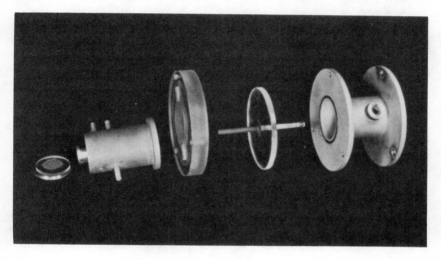

Fig. 5.39 Breakdown of the major components of the uncooled capillary discharge tube designed by Hunter [83].

cylinders sealed to allow water to flow between them. A hollow aluminum cylinder is seen protruding from the cathode and acts as the cathode electrode, allowing easy replacement. The aluminum insert has a few very fine holes to allow the discharge gas to leak into the discharge region. The anode (extreme right) is hollow and flanged to mate the spectrograph. A small window is inserted for visual observation. A Plexiglass flange is used to clamp components together and is grooved to locate the quartz disc such that the capillary lies on the axis of the spectrograph. Figure 5.40 illustrates the complete discharge lamp.

Typical spectra obtained with the Hunter-type discharge lamp are shown in Figs. 5.41 to 5.44. The discharge current was held constant at 400 mA.

The hydrogen spectrum (Fig. 5.41) is mainly due to molecular hydrogen. However, the atomic resonance lines Lyman-alpha and Lyman-beta can be seen with the Lyman-alpha line at 1215.7 Å dominating the spectrum.

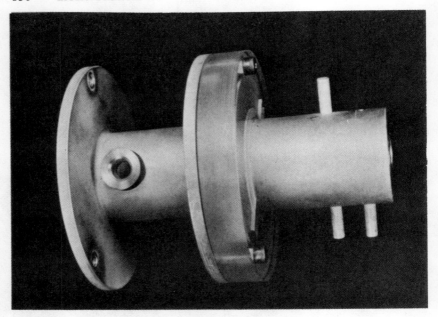

Fig. 5.40 Complete light source of the type shown in Fig. 5.39.

The band structure of hydrogen is very sharp, having the appearance of a line spectrum. This gives rise to the commonly used term "the many-lined molecular hydrogen spectrum." The band structure extends from about 850 to 1670 Å, at which point the H_2 continuum starts and extends to about 5000 Å. Like the continuum, the intensity of the band spectrum increases as the pressure increases. The intensities of the atomic lines are not so pressure dependent. Thus at very low pressures, the atomic resonance lines are more prominent than the band structure. With helium or argon in the discharge (and with a trace of hydrogen), the spectrum is almost entirely due to the atomic resonance lines. An excellent photoelectric atlas of the intense lines of the hydrogen band spectrum from 1025 to 1650 Å at a resolution of 0.1 Å has been prepared by Schubert and Hudson [84]. A photographic atlas has been prepared by Junkes et al. [85] covering the range 1260 to 1650 Å. In both of these atlases, the identification of wavelengths were made mainly from the published works of Herzberg [86] and Rosen [87].

The helium spectrum (Fig. 5.42) is dominated by the first resonance line of neutral helium at 584.334 Å. The second resonance line at 537.03 Å is down in intensity by about a factor of 60. A few weak impurity lines from oxygen, nitrogen, and hydrogen can be observed. The intensity of the

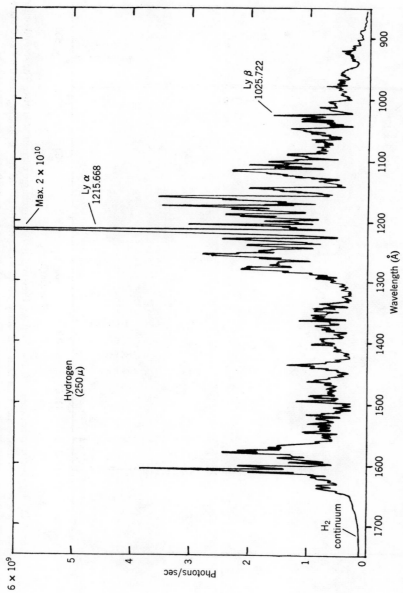

Fig. 5.41 H_2 spectrum between 850 and 1750 Å. Pressure, 250 μ; discharge current, 400 mA. No radiation is produced at wavelengths shorter than 800 Å.

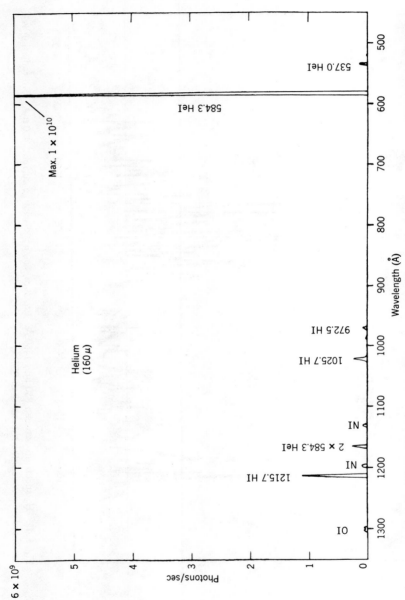

Fig. 5.42 He spectrum between 450 and 1350 Å. Pressure, 160 μ; discharge current, 400 mA. The second order 584 Å line is down in intensity by a factor of thirty compared to the first order line indicating that the stated blaze at 600 Å for the grating is correct. Impurity lines from O, N, and H are present.

584 Å line continually increases with pressure (at least over the range 60 to 600 microns). On the other hand, the intensity of the 537 Å line, although not so sensitive to pressure, tends to decrease as the pressure increases. The abnormal weakness of the 537 Å line relative to 584 Å and its decrease with increasing pressure can probably be explained by the fact that a collisional process takes place before the excited $3\,^1P$ state can radiate, such as

$$He^* + He \rightarrow He_2^+ + e. \qquad (5.23)$$

The above reaction is known to occur readily and has an onset energy at 23 eV corresponding to the $3\,^1P$ level of helium [88]. This process occurs

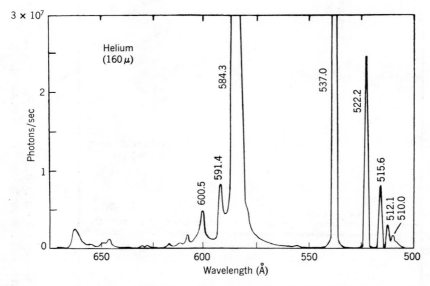

Fig. 5.43 He spectrum between 475 and 675 Å showing the principal resonance series and the two forbidden transitions at 591.4 and 600.5 Å.

for all of the resonant states with $n \geq 3$. Further evidence for the above reaction has been given by Teter et al. [89]. They have observed that the visible radiation originating from energy levels of principal quantum number $n = 3$ in the helium positive column shows a pressure dependent loss process. Although the intensities of the higher members of the resonance series are much weaker than the 584 Å line, they are still sufficiently intense to be useful. Figure 5.43 shows the region 500 to 650 Å of Fig. 5.42 greatly amplified. The forbidden transition from the $2\,^1S$ state to the ground $1\,^1S$ state radiating 600.5 Å can be seen, although it is

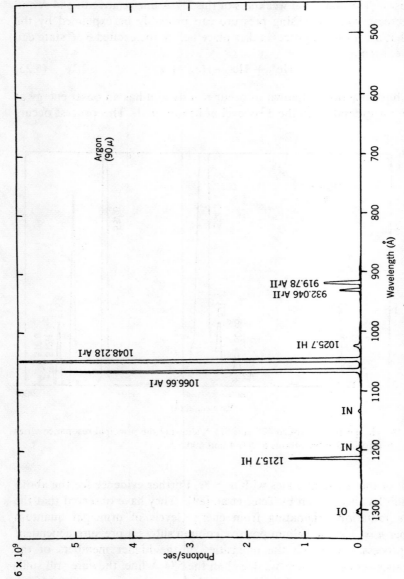

Fig. 5.44 Ar spectrum between 500 and 1350 Å. Pressure, 90 μ; discharge current, 400 mA. The resonance lines of Ar II can be seen as well as impurity lines from O, N, and H.

142

approximately 2000 times less intense than the 584 Å line. The inter-combination line at 591.4 Å $(2\ ^3P - 1\ ^1S)$ is also observed. The remaining lines identified in Fig. 5.43 represent the resonance series $n\ ^1P - 1\ ^1S$, $n = 2, 3, \ldots$.

The argon spectrum is shown in Fig. 5.44. The major lines in the spectrum are the Ar I resonance lines at 1048.2 and 1066.7 Å, the Lyman-α impurity line at 1215.7 Å, and the resonance lines 919.8 and 932.0 Å from Ar II. Other weak impurity lines from O, N, and H are also present. If the discharge current is kept constant and the argon pressure decreases, the

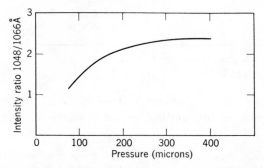

Fig. 5.45 Ratio of the intensities of the 1048 and 1066 Å resonance lines of Ar I as a function of pressure. Discharge current, 400 mA.

intensity of the Ar II lines increases. The relative intensity of the 1066.7 Å line to the *raie ultime* line at 1048 Å varies with pressure. As shown in Fig. 5.45, they are of comparable intensity at low pressure; however, the 1048 Å line is about 2.5 times more intense at higher pressures.

In krypton, the radiation is mainly due to the 1164.9 and 1235.8 Å resonance lines; in xenon, to the 1295.6 and 1469.6 Å resonance lines; and in neon, to the 735.9 and 743.7 Å resonance lines. As in helium, the higher members of the resonance series are either absent or very weak due to collisional processes such as described by (5.23).

When O_2 or N_2 are used in a glow discharge, the atomic resonance lines are excited; however, in the case of N_2, a molecular band spectrum is also observed extending from 900 to 1800 Å. This band spectrum is not so intense or so useful as the H_2 "many-lined" spectrum. The intensity of the atomic lines are about a factor of ten less intense than those of the rare gases. The most intense of the atomic lines observed are listed in Table 5.4.

In all cases, the intensity of each spectrum shown above increased almost linearly with the discharge current. The detector used to record the traces was an EMI 9514B photomultiplier sensitized to vacuum uv radiation with sodium salicylate. The actual spectral energy distribution

Table 5.4 Wavelengths of the most Intense Atomic Lines of O and N as Produced in a dc Glow Discharge[a]

O I $\lambda(\text{Å})$	N I $\lambda(\text{Å})$
1302.168*	1134.981*
1304.858	1134.415
1306.029	1134.166
	1199.550
	1200.224
	1200.710

* Indicates the *raie ultime*.
[a] See Table 10.3.

shown here is, of course, influenced by the individual grating used. In this case, a 1200 line per mm grating, platinum coated and blazed for 600 Å, was used.

Hollow Cathode. Another version of the glow discharge tube utilizes the radiation emitted from the negative glow. This is achieved by using a hollow cathode closed at one end. As the gas pressure is reduced, the discharge retreats within the cathode until the negative glow fills the hollow cathode. At this pressure, the electron mean free path is approximately equal to the cathode diameter. As mentioned earlier, the negative glow produces a spark spectrum in addition to an arc spectrum. Thus, the hollow cathode discharge tube is frequently used to produce radiation from singly ionized atoms (usually from the metal comprising the cathode). An inert gas is normally used for the discharge. This type of light source is particularly valuable for producing sharp spectral lines and is often used for measurements of standard wavelengths. The hollow cathode tube was originally used by Paschen [90] and Schüler [91] and is sometimes referred to as a Schüler lamp. Figure 5.46 shows a simple design suitable for use at low power levels. Newburgh et al. [92] have constructed a more rugged tube using a tantalum cathode with an internal diameter of 11.7 mm. The cathode is located at the center of a cylindrical water-cooled anode. They reported that the 304 He II line was more intense than the 584 He I line. However, the measurements were made with a grazing incidence monochromator whose efficiency for the production of those two lines was unknown.

More highly ionized spark lines can be produced in a Penning discharge tube (P.I.G.). Deslattes et al. [93] observed lines from Ne III and Ne IV

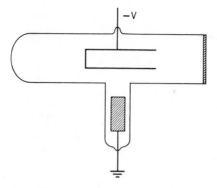

Fig. 5.46   Hollow cathode discharge tube.

between 200 and 360 Å when operating a P.I.G. discharge at 1 A and with a magnetic field of 1 kG. A major difficulty with this type of source, when operated at high power, is the rapid erosion of the cathode by the positive ion beam. Deslattes et al. [93] found that the erosion rate could be as high as 1 mm per h.

Since the intensity of the vacuum uv radiation varies with the gas pressure and the discharge current, it is important to keep both of these parameters constant in order to maintain the light output constant. A suitable circuit to regulate the current in a glow discharge has been described by Scouler and Mills [94] and is shown in Fig. 5.47. This circuit limits the maximum current to 300 mA since this is the maximum rating of the triode 6336. However, with two such tubes in parallel, the maximum current to the discharge tube can be increased to 600 mA. Using this circuit, Scouler and Mills have found that the discharge current did not vary by more than ± 0.1 per cent over a period of one day. It should be noted, however, that by simply using a regulated high voltage supply and allowing a suitable warming up period for the lamp, the intensity variations are generally only a few per cent.

Hot Filament Arc Discharge

In the preceding section it was noted that in a gas discharge, a transition between the glow discharge and an arc discharge takes place at some critical value of the discharge current. The arc discharge was characterized as a high-current (several amperes), low-voltage discharge. To provide the necessary high current to maintain the arc, severe ionic bombardment of the cathode is necessary, and considerable erosion occurs. A more simple and stable arc can be provided by using a separately heated filament to supply the necessary electrons. Several such lamps have been described in

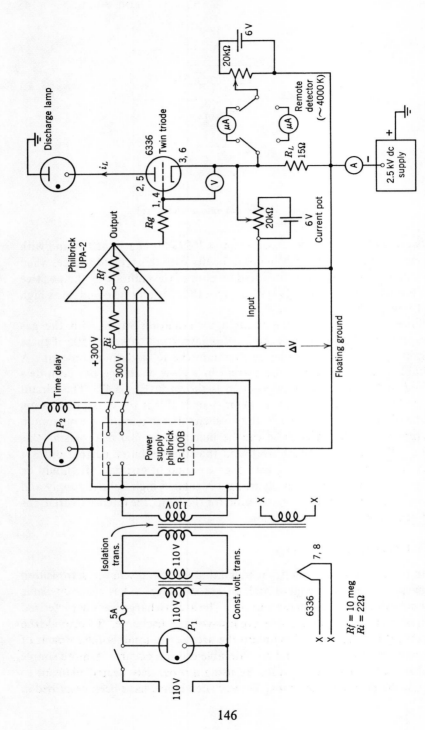

Fig. 5.47 Schematic of current regulator for a discharge lamp (courtesy W. J. Scouler and E. D. Mills [94]).

146

the literature [95–98]. The arc discharge described here is based on the design used by Hartman [96,97]. Figure 5.48 shows the lamp assembly. The filament is a helically coiled nickel ribbon dipped in barium carbonate and activated in a hydrogen atmosphere. The activation process consists of raising the temperature of the filament slowly in hydrogen at a pressure of about 1 torr or less. The current through the filament is increased from zero to about 15 A over a period of 30 min. A trigger pin is inserted in the

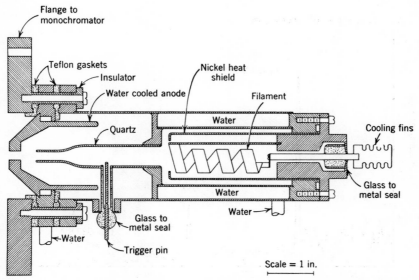

Fig. 5.48 Hot filament discharge lamp based on a design by Hartman [96, 97].

quartz discharge tube to provide a means for initiating the arc. A Tesla coil is suitable to trigger the discharge. The filament draws approximately 12 A at 4 V, while the voltage across the arc is about 90 V when an arc current of 3 A is being passed in hydrogen. Thus, the light source can be operated at 270 W. The dc glow discharge described on pp. 137 to 143 was operated at approximately 240 W.

Although the arc discharge produces a strong hydrogen molecular spectrum, the atomic resonance lines appeared clearly above the weaker molecular bands. By using a mixture of hydrogen and helium in the discharge (approximately 25 percent H_2), the radiation is nearly monochromatic in the Lyman-α line. Actually, the intensity of the atomic lines in the pure H_2 discharge did not change appreciably when helium was added; however, the intensities of the molecular lines were greatly reduced. Figure 5.49 shows the Lyman series of atomic hydrogen when a H_2-He mixture was used in the arc discharge.

The hot filament arc discharge in H_2 produces a more intense spectrum than the cold cathode glow discharge. The intensity ratio is only a factor of 2 or 3. The ratio of atomic to molecular line intensities is much greater for the hot filament arc discharge. One advantage in using the hot filament is that a discharge current of several amperes can be created at low voltages, and these supplies are more easily regulated and are less expensive than

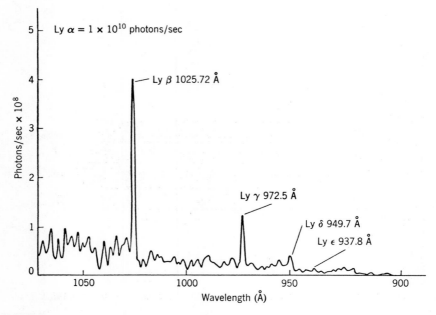

Fig. 5.49 Lyman series of atomic hydrogen obtained in a H_2-He mixture with the hot filament discharge lamp. The intensity of the Lyman-α line is given at the top of the figure.

regulated high voltage supplies. A prime disadvantage is the need to renew filaments periodically; however, with some experience, this does not present a major obstacle.

The circuit for a simple 150 V, 5 A power supply is shown in Fig. 5.50. The supply is current stabilized by simply inserting an amperite ballast tube in series with the load. Figure 5.51 shows the current-voltage characteristics of a typical tube. The power supply shown in Fig. 5.50 was constructed with ten amperite 3-38A ballast tubes in parallel in order to provide current stabilization from 0.3 A to 3 A in steps of 0.3 A by simply switching in the number of tubes required to provide the desired arc current.

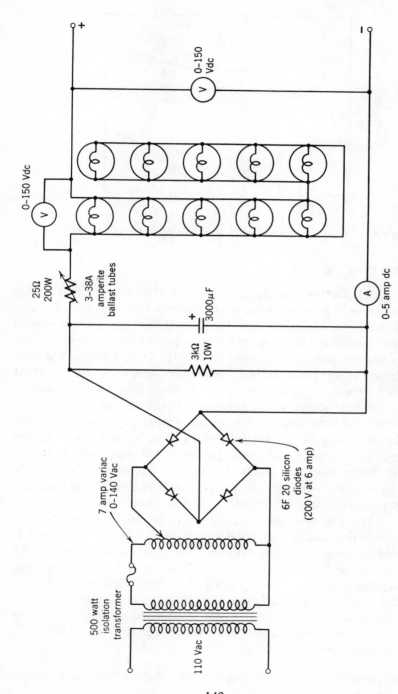

Fig. 5.50 Current regulated arc power supply for 0.3 to 3 A.

149

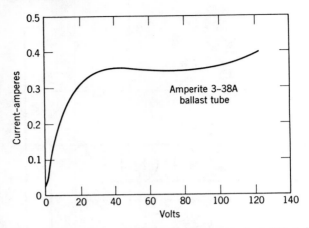

Fig. 5.51   Current-voltage characteristic of an amperite 3-38A ballast tube.

Duoplasmatron

The duoplasmatron was developed about 1950 to provide a highly efficient source of protons. After the publication [99] of its main features in 1956, a number of variations of the original instrument were designed and used as ion or electron sources for such applications as accelerator ion sources and ion propulsion devices [100–103]. More recently, the radiation from the highly concentrated plasma has been used as an excellent source of vacuum uv radiation [104]. It differs from the conventional hot filament light source in that a magnetic field is applied that constricts the discharge to form a narrow plasma beam along the axis of the source. The plasma density on the axis near the anode increases quickly with the magnetic field strength and, after passing a flat maximum, decreases slightly. In practice, this means that the duoplasmatron must operate above a minimum magnetic field strength if a maximum plasma density is to be provided on the axis. A sectional view of a duoplasmatron is shown in Fig. 5.52. The magnetic field can be produced by a conventional electromagnetic; however, in the design shown in Fig. 5.52, the field is produced by three permanent magnets. The magnets are formed in the shape of rings and are magnetized in the direction of their common axes. The magnetic material is a highly oriented barium ferrite that has a high electrical resistance. The trade name of this material is Index V.*

The magnetic flux goes from one pole of the magnet through the source flange and the slit holder, across the gap through the wall of the baffle electrode, and then through the cover plate to the other pole of the magnet.

* Index V, model number F-5911, produced by Indiana Steel Products, Valparaiso, Ind.

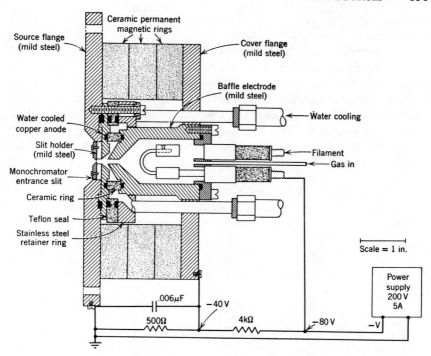

Fig. 5.52 Duoplasmatron light source. The voltage distribution shown is typical under discharge conditions in hydrogen.

The slit holder must be made from mild steel to concentrate the magnetic flux between the apex of the baffle electrode and the slit holder. For the particular magnets used in Fig. 5.52, a magnetic field of 7000 oe was created. The magnetic field presses the slit holder against the water-cooled copper anode. O-rings are used throughout for sealing.

The entrance slit of a monochromator can be removed and the duoplasmatron located such that the position of the entrance slit coincides with the slit holder of the duoplasmatron. Slits made from copper and attached firmly to the slit holder rapidly conduct away the heat created by the large arc currents. The orifice of the slit holder is about 3 mm in diameter, thus providing a maximum slit height of 3 mm. The width of the slit was 50 μ.

The filament is constructed of platinum mesh wire. After cleaning the filament in nitric acid and distilled water, the filament is dipped into a suspension of barium carbonate and then allowed to dry. The activation of the filament can be performed in a vacuum by slowly increasing the filament current from about 5 to 15 A over a period of one hour. In operation, the filament requires about 12 A to provide an adequate electron emission. As long as the filament is not exposed to the air, no subsequent activation is necessary.

The electrical connections are also shown in Fig. 5.52. The arc power supply should be capable of providing a maximum voltage of 200 V at about 3 A. The resistors shown in the diagram allow most of the applied voltage to appear between the filament and the baffle electrode before the discharge commences. Once the discharge has started, the potential difference between the electrodes is dependent on the discharge characteristics. The voltage distribution shown in Fig. 5.52 is typical under discharge conditions in hydrogen. Once an arc is established, the total

Table 5.5 Typical Electrode Potentials in a Duoplasmatron for Different Gases at an Approximate Pressure of 350 Microns

Gas	Arc Current (A)	Filament Potential (V)	Baffle Electrode Potential (V)	Anode Potential (V)
Ar	1	−55	−25	0
Kr	3	−31	−24	0
Ar	3	−45	−36	0
Ne	3	−66	−60	0
He	3	−82	−70	0

voltage drop across the duoplasmatron, although dependent on the nature and the pressure of the gas, does not vary much with the arc current. However, the potential of the baffle electrode increases with the arc current Table 5.5 shows some typical electrode potentials for various gases at a pressure of about 350 μ. The potential drop between the baffle electrode and the anode also increases as a function of the ionization potential of the rare gas used. Thus helium has the largest potential drop.

When a discharge is started, the gas pressure in the duoplasmatron increases while the pressure in the monochromator decreases. With an arc current of 3 A, the pressure in the duoplasmatron practically doubles. This is apparently caused by the intense ionization in the vicinity of the entrance slit. The incoming gas is ionized and forced back to the filament by the electric field impeding the flow of neutral gas through the slit into the monochromator.

The radiation emitted from duplasmatrons differs from that of glow discharges and hot filament discharges in that duoplasmatrons produce strong resonance radiation not only from neutral atoms, but also from singly ionized atoms. At large arc currents, the second spectrum is often more intense than the first. Weaker lines from doubly ionized species can also be seen. The intensity of a line increases approximately linearly with the arc current; however, the intensity of the ion spectrum increases more rapidly than that of the neutral atoms. At greater currents, the intensity of the radiation tends to saturate with saturation occurring first for the neutral atom spectrum. Typical spectra obtained from argon and neon with an arc current of 3 A are shown in Figs. 5.53a, b and 5.54a, b. With the exception of some weak impurity lines from N I and H I, the argon spectrum is due primarily to Ar II lines. A direct comparison can be made with the argon spectrum in Fig. 5.44 produced by a glow discharge. Although the glow discharge does produce the main resonance lines of Ar II, no lines of useful intensity appear at wavelengths shorter than 900 Å, whereas the duoplasmatron produces lines of useful intensity down to 450 Å. Although many of the lines in Fig. 5.53 appear weak relative to the neighboring strong lines, it should be noted that intensities as weak as 10^7 photons per sec are extremely useful for many applications, especially if the equivalent noise level is considerably lower. For example, note the group of Ne II lines shown in Fig. 5.54 (a) between 310 and 380 Å, and note the enlargement of this area in Fig. 5.54 (b). At wavelengths longer than 500 Å, the neon spectrum is dominated by the two resonance lines at 735.89 and 743.70 Å. The intensities of these two lines were 15×10^9 and 4.5×10^9 photons per sec, respectively.

In all cases studied, the duoplasmatron produces a strong ion spectrum. Even in the case of helium, the 304 Å He II line appeared with an intensity of 3×10^7 photons per sec. Although the 584 Å He I resonance line had an intensity of approximately 10^{10} photons per sec, the true intensity ratio of the 584 to the 304 Å line emerging from the duoplasmatron is probably not so great as the ratio emerging from the exit slit of the monochromator, since the grating was blazed for 600 Å. In the case of krypton, the Kr II spectrum is the most intense, producing a dense number of lines between 600 and 965 Å with intensities lying between 10^9 and 10^{10} photons per sec. It is expected that a similar spectrum exists for xenon. When nitrogen is used, there is very little band radiation as is observed with the glow discharge; instead, the spectrum again consists primarily of N I and N II lines. The H_2 discharge produces the familiar many-lined molecular spectrum and H_2 continuum; however, the intensity of the molecular spectrum is much less than that produced in a glow discharge. On the other hand, the intensities of the atomic lines are much greater. As the arc current was increased, the intensity of the radiation increased almost linearly with the H_2 molecular lines increasing at a slower rate than the atomic lines. This result can be correlated with the results of the plasma beam composition reported by Moak et al. [100], namely, that the number of H^+ ions increased more rapidly than the number of H_2^+ ions as the arc current increased.

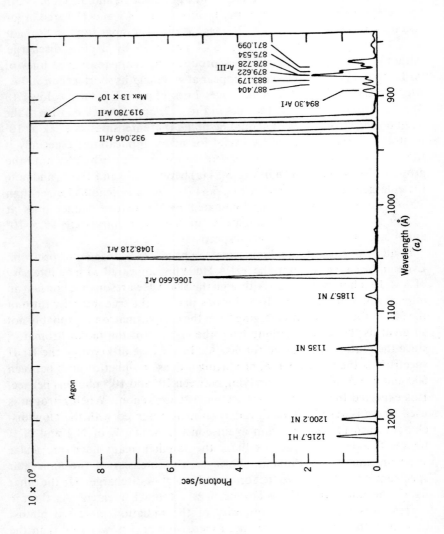

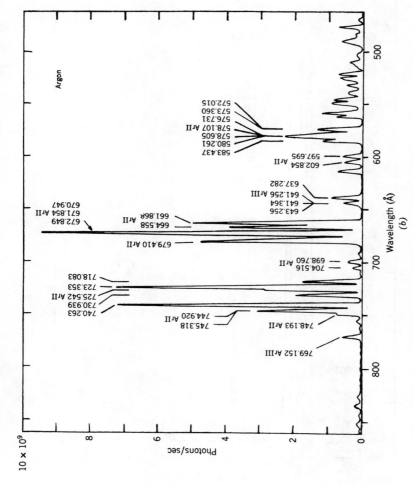

Fig. 5.53 Ar spectrum between 450 and 1250 Å produced by a duoplasmatron. Pressure, 380 μ; arc current, 3 A.

155

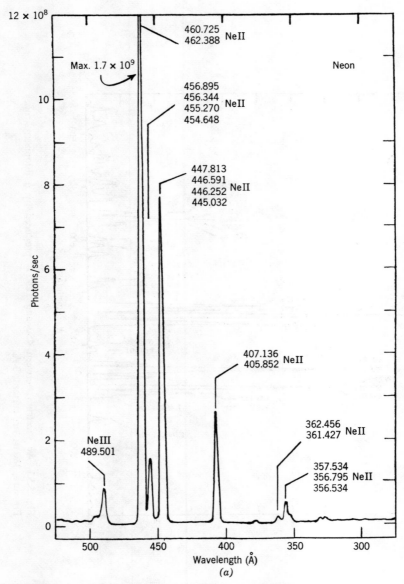

Fig. 5.54 (*a*) Ne spectrum between 300 and 500 Å produced by a duoplasmatron.

156

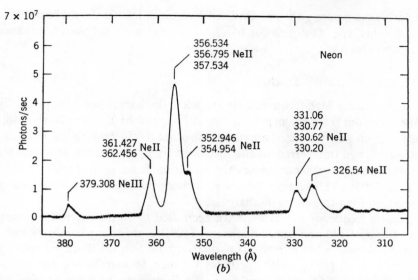

Fig. 5.54 (b) Enlargement of Ne spectrum between 310 and 380 Å. Pressure, 450 μ; arc current, 3.3 A.

Braams et al. [103] have studied the composition of the rare gas ions extracted from the duoplasmatron. They observed that the relative intensity of a given ion varied with the pressure of the gas in the duoplasmatron. Usually, the ion intensity increased to a maximum and then decreased as the pressure was increased. Table 5.6 lists the pressure at

Table 5.6 Pressure at the Maximum
Yield for a Given Ion Type[a]

Ion	Pressure at max. yield (torr)
He^+	0.700
Ne^+	0.250
Ar^+	0.250
Ar^{2+}	0.110
Kr^+	0.175
Kr^{2+}	0.065
Kr^{3+}	0.065
Xe^+	0.120
Xe^{2+}	0.070
Xe^{3+}	0.070

[a] C. M. Braams, P. Zieske, and M. J. Kofoid, *Rev. Sci. Instr.* **36**, 1411 (1965).

which a maximum yield was obtained for a given ion. It should be possible to use this type of information to adjust the pressure conditions to enhance a given spectrum.

Microwave and RF Discharge

Continuous high-frequency electrodeless discharges produce primarily the spectrum from the neutral atom, although at high power levels, radiation from singly ionized atoms can also occur. Increasing the instantaneous power transferred to the discharge by operating the hf oscillator intermittently further increases the radiation from more highly ionized atoms [105]. This section, however, is concerned only with continuously operated high-frequency discharges.

Radio frequency oscillators have been used to excite electrodeless ring discharges [106] and linear discharges [107] to produce vacuum uv radiation. A simple but effective rf oscillator, based on the designed used by Garton et al. [107], which oscillates between 16 and 20 Mc per sec, is shown in Fig. 5.55. The coupling to the discharge tube can be optimized

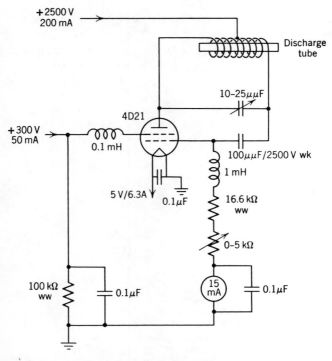

Fig. 5.55 Radio-frequency oscillator circuit, 16 to 20 Mc per sec.

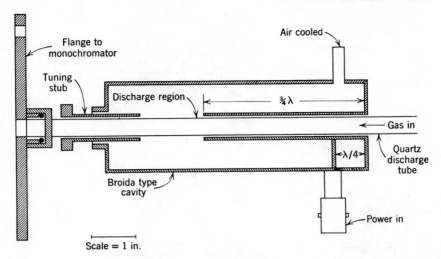

Fig. 5.56 Microwave light source and cavity.

using a variable capacitor in the oscillating circuit. The theoretical power output of this oscillator is 150 W. The spectra produced by an rf discharge are similar to those produced at microwave frequencies. The availability of medical microtherm generators,* operating at 2450 Mc per sec and 125 W, and of commercial microwave cavities† has popularized the use of microwave frequencies to excite discharges. The main disadvantage of the simple medical microtherm units is their lack of power regulation and their limitation on the output power level. The Jarrel-Ash Company has produced a commercial microwave generator which is regulated and has an output of 200 W. Unregulated supplies producing up to 800 W are manufactured by the Raytheon Company.

The use of microwave frequencies to excite vacuum uv radiation has been described by numerous authors [108–116]. Although many methods for coupling the microwave power into the discharge have been described, the most efficient method is a properly constructed and tuned cavity. A thorough study of the various modes of microwave coupling to discharges has been carried out by Fehsenfeld et al. [117]. Figure 5.56 shows a frequently used cavity surrounding a quartz discharge tube. The tuning stub must be adjusted to provide maximum coupling. The most efficient cavity designed by Fehsenfeld et al. is shown in Fig. 5.57. With a tunable stub and a coupling slider, this cavity reflects less than one per cent of the

* Burdick Company, Milton, Wisconsin.
† Opthos Instrument Company, Rockville, Maryland; and G. B. L. Associates, Alexandria, Virginia.

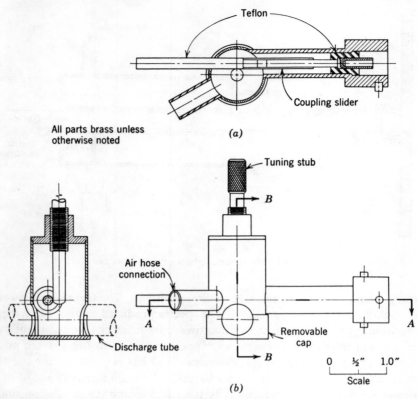

Fig. 5.57 Microwave cavity. This design reflects less than 1 per cent of the input power when properly tuned (courtesy F. C. Fehsenfeld et al. [117]).

incident power over a pressure range of 0.1 to 100 torr. An added advantage is the removable cap allowing the cavity to be easily clipped onto the discharge tube without breaking the tubing.

The spectra produced in microwave discharges are similar to those of the dc glow discharge and hot filament arc discharges although probably slightly weaker because of the lower power levels available with the medical-type generators. Like the hot filament arc and duoplasmatron, the microwave discharge in hydrogen favors the atomic spectra at the expense of the molecular lines. In a hydrogen-helium mixture, the microwave discharge produces perhaps the most monochromatic source of Lyman-alpha radiation. Figure 5.58 shows a spectrum obtained with a H_2-He mixture containing approximately 25 per cent H_2 and a total gas pressure of 200 microns. Figure 5.59 shows the higher members of the Lyman series obtained with a mixture containing about 2 per cent H_2 at a total pressure of 500 microns. Okabe [115] has obtained similar results using a

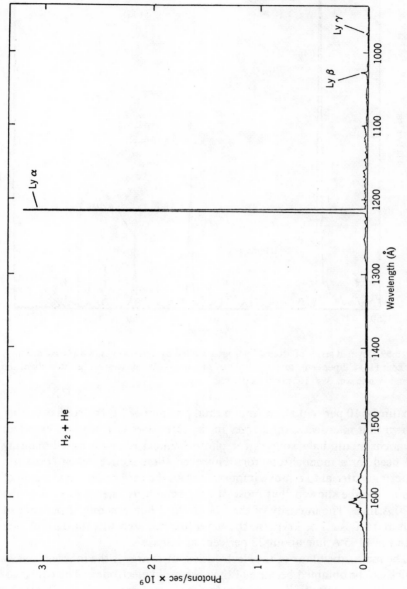

Fig. 5.58 Hydrogen spectrum excited by microwaves in a H_2-He mixture.

161

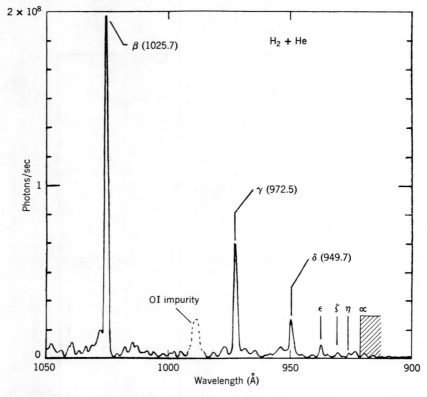

Fig. 5.59 Lyman series of atomic hydrogen excited by microwaves in a H_2-He mixture. Per cent $H_2 \sim 2$ per cent, total pressure of mixture $\sim$500 microns. The intensity of the Lyman-α line was 3×10^9 photons per sec.

mixture of 10 per cent H_2 in Ar at a total pressure of 1 torr. He has further shown that microwave discharges in the rare gases can produce excellent monochromatic light sources for photochemical research by eliminating the need for a monochromator. However, these sources must contain a gettering material to remove impurities from the rare gases. In the case of xenon, Okabe showed that most of the vacuum uv energy existed in the 1470 Å line. The intensity of the 1295 Å Xe I line was only 2 per cent of that at 1470 Å. For krypton, the main line radiated was the 1236 Å line with the 1165 Å line about 22 per cent as intense.

The main advantage of the electrodeless discharge is the inherent purity which can be obtained because of the absence of electrodes. The construction of the source itself is extremely simple and inexpensive. Quartz or Pyrex tubing can be used, however, quartz has a higher melting point and has a lower loss factor than Pyrex.

Condensed Spark Discharges

The vacuum spark or hot spark was first used by Millikan [118] to study the spectra of metals in the extreme ultraviolet. He discovered that when extremely high potentials were applied between two electrodes a millimeter or less apart in a high vacuum (10^{-5} to 10^{-6} torr), a brilliant spark occurred between them. To provide the necessary current density, a capacitor was connected across the two electrodes. Field strengths of the order of 150 kV per mm were used to initiate the discharges. With this source, Millikan was able to push the short wavelength limit of the vacuum uv region down to 140 Å.

Lyman [119] used a light source similar in design to that of Millikan but introduced a gas at low pressure which produced sparks at reduced voltages. The radiation emitted in this case was characteristic of the gases used in the light source.

The sources described above are the basis of the two types of condensed spark discharges used at present, namely, a high-vacuum spark discharge and a low-pressure spark discharge.

With the Millikan spark discharge, the electrodes are rapidly worn away and the sputtered material clogs the spectrograph slit. Vodar and Astoin [120] discovered that a high vacuum spark discharge could be produced at lower voltages and with less sputtering of electrode material if the electrodes were separated by, but in good contact with, an insulator. The spark discharge occurred on the surface of the insulator giving rise to the name "sliding spark discharge." Vodar and Astoin used a capacitor of 0.03 to 0.1 μF charged to 30 kV. The radiation emitted was characteristic of the metal of the electrodes and also of the insulating material. The separation of the electrodes can be much greater in the sliding spark discharge than with the Millikan source. Many descriptions and modifications of the sliding spark light source have been published [121–131]. Figure 5.60 shows a typical arrangement of the electrodes and insulator suggested by Edlén and used by Bockasten [121]. The ceramic insulator has a slot cut in its side to allow observation of the radiation. For proper operation, the electrodes must be pressed firmly in contact with the ceramic. A suitable electrical circuit is also shown in Fig. 5.60. An external spark gap is used in series with the vacuum sliding spark to insure that the discharge breaks down at a constant potential since the ceramic tends to become coated slightly with electrode material causing a reduction in its surface resistance. Bockasten has observed that an added advantage of the sliding spark source is the fact that there is much less Stark broadening of the spectral lines compared with those produced by the hot spark discharge. To produce radiation from extremely highly ionized atoms, the

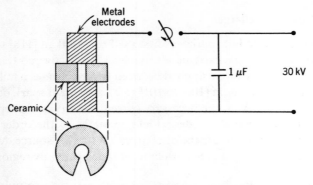

Fig. 5.60 Arrangement of the electrodes and insulator in the sliding spark light source. The sparks slide along the internal diameter of the ceramic insulator.

inductance of the circuit should be very low or the voltage extremely high. In earlier research low-inductance capacitors were not available thus very high voltages were necessary to produce the desired high degree of ionization. Inductance can be purposefully introduced in series with the light source to help identify the state of ionization producing unknown spectral lines, since radiation from the most highly ionized atoms disappears first as the series inductance is increased [121,132]. The main usefulness of high vacuum spark discharges is in the study of the emission spectra of the electrode material and for research applications where no trace of light source gas leaking into the experimental chamber can be tolerated.

In order to study the spectra from atoms ionized to a much greater degree than perhaps can be obtained from the vacuum sparks such sources as the Zeta pinch and other thermonuclear plasma devices have proved suitable [133–136]. These spectra have been extremely useful in producing comparison spectra to identify unknown lines in the solar spectrum [137]. The sun itself, of course, is an excellent source of vacuum uv radiation.

Perhaps the most useful source of vacuum uv radiation for general use below 1000 Å is the condensed spark discharge in a low pressure gas. The discharge is confined in a capillary a few centimeters long and a few millimeters in diameter. This type of source was originally used by Lyman and later by Paul and others [138]. In the mode of operation used by those earlier researchers the source was quite adequate to provide emission spectra from highly ionized gas atoms; however, it was not suitable for many quantitative measurements where the intensity of the emission must stay constant from spark to spark over prolonged periods. Weissler and associates pioneered the development of a low-pressure spark source which would meet those more stringent requirements. The first description of such a source appeared in 1952 by Po Lee and Weissler [139]. In this source, no external spark gap was used, and an ac voltage was supplied from a high voltage transformer. An oscilloscope monitored the instantaneous potential across the capacitor, while the gas pressure was adjusted until the oscilloscope pattern was constant. With this type of source, the breakdown potential is a function of the gas pressure; consequently, radiation from highly ionized states may not be produced. They observed that by varying the gas pressure and hence breakdown potential, that only N I lines appeared at 1130 V, N II lines at 1840 V, N III lines at 2200 V, and N IV lines at 4000 V. To avoid the breakdown potential

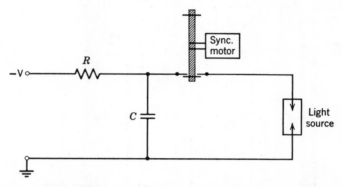

Fig. 5.61 Rotary spark gap and circuit for the condensed spark discharge. Repetition frequency 60 pulses per sec. $R = 12.5 \ k\Omega$, $C = 0.25 \ \mu F$, V = 6 to 8 kV.

dependence on the pressure, Wainfan et al. [140] used an external spark gap in series with the light source. This gap was triggered mechanically by a rotating electrode, as shown in Fig. 5.61. A synchronous motor must be used to insure that the gap is triggered at the same point on the RC charging curve for every spark. Any number of electrodes may be installed in the rotating disc; however, a practical limit is set by the amount of power which can be dissipated in the light source. Typically two or three rotating electrodes are used providing 60 to 90 pulses per sec. These earlier sources operated at about 15 kV with a capacitor of about 0.04 μF. However, the inductance of the circuit and capacitors was not minimized. Using the special low inductance capacitors currently available, more light intensity can be achieved for the same input power.

Many light source designs have been reported in the literature [83,141–144]. However, any condensed spark discharge in a low pressure gas produces essentially the same spectra. For example, the use of a simple ceramic high voltage feed-through has served successfully as a light source. Of course, not all sources are as stable or as rugged as others. A design used by the author for several years and based on that used by Weissler

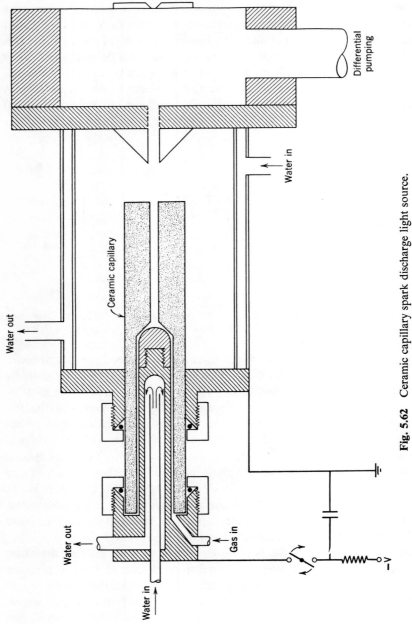

Fig. 5.62 Ceramic capillary spark discharge light source.

Differential pumping

Water in

Ceramic capillary

Water out

Water out

Gas in

Water in

$-V$

166

and associates is shown in Fig. 5.62. The important features in the design are as follows. The anode and cathode should be water cooled, and the ceramic material should be capable of withstanding the high temperature generated in the capillary and should be impervious to gases. Finally, an expansion volume should be provided to allow the shock wave generated by the discharge to expand and dissipate its energy. Otherwise, small particles of dust, ceramic, or electrode material may clog the spectrograph slits or pass through the slits and damage the grating. A set of four to six triangular fins has proved helpful in preventing charged material from clogging the slits [28]. The cathode shown in Fig. 5.62 has a demountable tip which must be occasionally replaced. The electrodes are generally constructed from aluminum; however, Rustgi [142] and Weissler [143] have used a copper-tungsten alloy (Elkonite) for their electrodes. They have also used a new material made from boron nitride that has a prime advantage: it can be machined to any desired shape without subsequent "firing." Boron nitride can withstand the necessary temperatures and has excellent thermal shock properties. It has some disadvantages, however, in that it reacts with a humid atmosphere and is attacked by oxygen if this gas is used in the discharge for extended periods. The cleanest ceramic to use contains aluminum oxide. Several varieties are manufactured, and their nature depends on the percentage of aluminum oxide and other additives. Alsimag 614, the trade name of an alumina-silicon-magnesium mixture, has been used for many years without requiring attention and shows virtually no wear. Its only disadvantage is its susceptibility to fracture when the discharge is operated at very high power levels.

The optimum length and diameter of the capillary has never been determined. However, it has been observed that capillaries less than 3 cm long produce less light than capillaries 5 cm long. Values used for the source described here are 5 cm (length) and 3 to 4 mm (diameter).

Some common problems encountered with the repetitive spark discharge should be mentioned. For example, at certain gas pressures, there is a tendency for a discharge to take place in the gas inlet tube. This causes instabilities in the light output. Although the pressure in the source can be adjusted to prevent this discharge, it is more desirable to eliminate it by using a plug of glass wool in the gas line as close as possible to the light source. Another problem is the electrical noise which can be generated from the source and the external spark gap. This can often be eliminated by adequate shielding and proper grounding procedures. Multiple grounds should be eliminated to prevent the formation of radiating ground loops. Preferably, the light source should be insulated from the monochromator in such a way that no part of the discharge crosses over to the monochromator.

The source shown in Fig. 5.62 is normally operated with a 0.25 μF low inductance capacitor charged to 6 kV. A resistor bank with a total resistance of 12.5 kΩ is used in the charging circuit. With these parameters, an average current of 100 mA is drawn from the regulated dc power supply. The rotating switch provides 60 pulses per sec. Thus the charging time of the capacitor between pulses is about 5.5 times RC. For the source to operate with a constant intensity, it is advisable not to switch during a rapidly rising portion of the charging curve; thus a charging time equivalent to several times the time constant RC should be used. The switching electrodes are best made from drawn tungsten as distinct from the pressed form. A tungsten rod 3 mm in diameter has been used. The rotating electrodes are placed about 25 cm apart. These dimensions are not critical, but are supplied to indicate values which have provided successful operation. The gap separation between the stationary and the rotating electrodes is made as small as possible. A quartz mercury "pen-ray" lamp has been used to irradiate the spark gap, and a stream of compressed air has been directed on the gap to provide increased stability and a constant light output. The electrical leads from the spark gap to the light source should be arranged to introduce minimum inductance to the circuit. A parallel plate transmission line or a special low inductance cable should be used.

The mechanical switch for the external gap can be replaced by a thyratron circuit as shown by Hunter [83]. This has the advantage of eliminating maintenance of the spark gap and of silent operation. In addition the repetition frequency can be easily varied. The main disadvantage is that the thyratron circuit requires more power input to produce the same light intensity as that given by the mechanical switch. This is possibly because of the increased inductance caused by the thyratron circuit. Hunter reports a total inductance of 1.1 μH with approximately 0.4 μH residing in the particular thyratron used (type 1257 hydrogen thyratron). The total inductance of the source and spark gap circuit shown in Fig. 5.62 was 0.47 μH. When the capacitor was charged to 6 kV and discharged through the light source, the current maximum was about 3000 A. In an LCR oscillating circuit, the current maximum i_{max} is approximately equal to $V_0\omega C$ or, more rigorously,

$$i_{max} = V_0\omega Ce^{-\alpha\tau}\left(1 + \frac{\log \text{dec}^2}{\pi^2}\right), \tag{5.24}$$

where V_0 is the initial voltage of the capacitor, τ the time taken for the current to reach a maximum value, and α is equal to twice the frequency times the logarithmic decrement. That is,

$$\alpha = 2f\log \text{dec}, \tag{5.25}$$

where

$$\log \text{dec} = \ln\left(\frac{i_1}{i_2}\right). \tag{5.26}$$

The ratio (i_1/i_2) represents the ratio of any two consecutive peaks in the oscillating current oscillograms. The total inductance of the oscillating circuit is given by

$$L = \frac{1}{\omega^2 C(1 + \log \text{dec}^2/\pi^2)}. \tag{5.27}$$

From (5.24), the maximum current is shown to be proportional to ω which in turn is approximately equal to $(1/LC)^{1/2}$. Thus i_{max} is inversely proportional to the square root of the total inductance; that is

$$i_{max} \propto \frac{1}{\sqrt{L}}. \tag{5.28}$$

Curves illustrating the instantaneous voltage, current, and light intensity as a function of time are shown in Fig. 5.63. The current and voltage are approximately 90° out of phase. The light intensity tends to follow the current, reaching a maximum each half cycle corresponding to a current maximum. If such oscillograms are studied for radiation originating from different stages of ionization, it is noticed that the radiation from N II, for

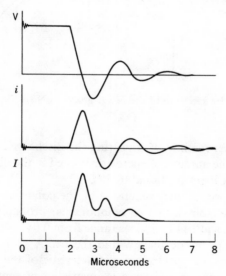

Fig. 5.63 Oscillograms of the instantaneous values of the voltage V, current i, and the light intensity I as a function of time from the start of the discharge. A jitter time of several microseconds occurs before the discharge starts.

example, has several more maxima than radiation from N V. This is because the oscillating voltage of the capacitor has fallen below that necessary to produce a high degree of ionization. With a thyratron as a switch, the discharge current flows only during the first half cycle because of the rectifying effect of the thyratron. The intensities of emission lines

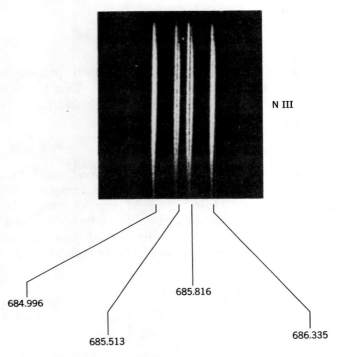

N III

684.996

685.816

685.513

686.335

Fig. 5.64 Third-order spectrum of the 685 Å group of N III illustrating the splitting of the lines.

produced by either method of triggering the discharge are approximately proportional to the amount of energy dissipated in the discharge. That is, the light intensity is proportional to V^2C.

Many of the spectral lines generated by the condensed spark discharge are broadened or split into two components, probably because of the Stark effect. The widths of the lines range from 0.04 to 0.1 Å. Figure 5.64 shows the splitting of the lines within the 685 Å group of N III [145]. Although the emission spectra are characteristic of the gases used in the light source, the oxygen spectrum is invariably present as an impurity. All the rare gases, O_2, and N_2 can be used in the light source to produce radiation from the respective ionized atoms. The neutral and singly ionized spectra are absent. The addition of a small amount of CO_2 to N_2 produces the ionized carbon spectrum. The spark spectrum of argon between 400 and 650 Å is shown in Fig. 5.65. Argon has an extremely dense line emission spectrum and is one of the best gases for general use between 400 and 1000 Å. Neon is particularly useful between 350 and 450 Å. Nitrogen produces more intense lines at shorter wavelengths. The shortest wavelength utilized with a Seya monochromator for quantitive research is the 209.270 N V line [146].

The intensity of the radiation is very sensitive to the pressure. Thus care must be taken to maintain a constant pressure within the light source. The intensity variations with pressure are different for lines originating from different stages of ionization. Figure 5.66 shows some typical results for Ar IV to Ar VIII as obtained by Schönheit [147]. This type of behavior has been found for all gases studied.

It is important to emphasize that the radiation emerging from the condensed spark discharge has a duration of only one to two microseconds. With an average intensity of about 10^{10} photons per sec radiated by a given line, the intensity per pulse can be very high and may cause saturation effects in a photomultiplier recording the spectrum. The current in the dynode resistor string should be about 1 to 10 mA with capacitors of the order of 0.01 μF across the last few stages. With this arrangement, no saturation effects have been observed even although anode peak currents of 20 mA have been produced by the multiplier. Of more serious concern, however, is the possible saturation of a dc microammeter by the photomultiplier. Although the radiation from the condensed spark discharge is pulsed, it can be readily recorded by a dc instrument when repetition rates of the order of 60 pulses per sec are used. The input resistance to a microammeter is generally very high, and along with the natural capacitance of the photomultiplier and connecting cables the pulsed signal is effectively integrated. However, in some microammeters, the electrometer tube is saturated by the sharp photomultiplier pulses. This can easily be corrected by using a simple integrating circuit between the photomultiplier anode and the microammeter input as show in Fig. 5.67.

For weak signals whose average intensity is comparable with the dark current noise produced by the photomultiplier, advantage can be taken of the pulsed nature of the source, since the instantaneous intensity of the weak signals is still greater than the intensity of the individual noise pulses. The noise is, however, present all the time; consequently, when the signals are averaged, the noise is dominant. To minimize the effect of noise, the detecting circuit is gated to receive a signal only during a discharge within the light source. A circuit due to Judge [148] is shown in Fig.

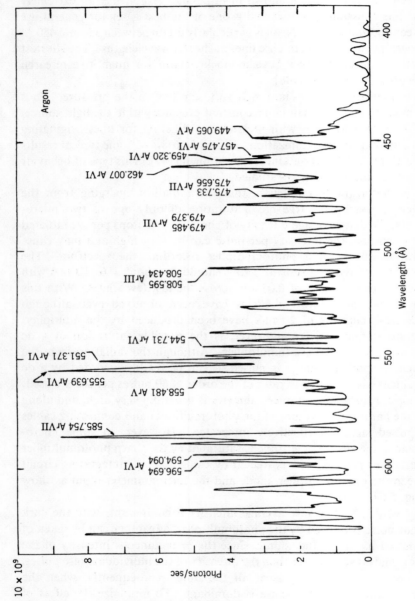

Fig. 5.65 Argon spectrum between 400 and 650 Å excited by a condensed spark discharge. Pressure ~100 μ; voltage, 5 kV; current, 74 mA.

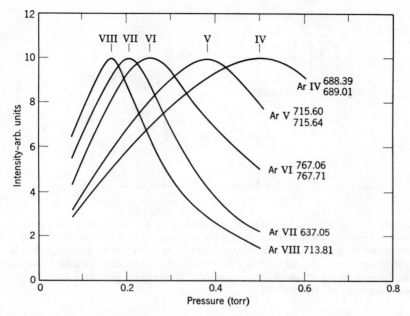

Fig. 5.66 The variation of intensity with pressure for lines from various stages of ionization. Discharge parameters were $C = 0.25\ \mu\text{F}$, $V = 6.3\ \text{kV}$, $L = 0.5\ \mu\text{H}$. The optimum pressure in microns to give a maximum intensity for a given line was $500\ \mu$ for Ar IV, $380\ \mu$ for Ar V, $250\ \mu$ for Ar VI, $200\ \mu$ for Ar VII, and $160\ \mu$ for Ar VIII (courtesy E. Schönheit [146]).

5.68. The function of the amplifiers is to amplify the output voltage to the point where they can pass through the diode FD 300 (the breakdown voltage is $\sim 0.5\ \text{V}$). The 2N417 transistor is biased such that it is continually conducting thereby short circuiting the input to the diode and micro-ammeter. The bias on the transistor is synchronized with the light source

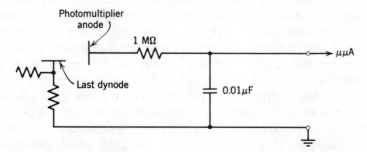

Fig. 5.67 Simple intgrating circuit to prevent saturation of the signals in the micro-ammeter.

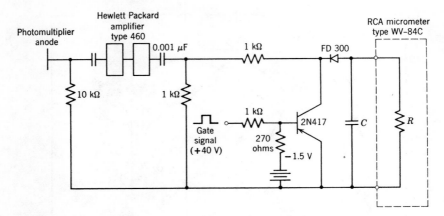

Fig. 5.68 Circuit for the synchronous detection of pulses enhancing the signal to noise ratio from a photomultiplier. When the transistor 2N417 is conducting, it has a resistance of 50 ohms; when it is nonconducting, its resistance to ground is 10 kΩ. Amplified pulses must exceed 0.5 V to overcome the diode barrier. The back resistance of the diode is about 10^9 ohms. The value of C is chosen to provide a suitable time constant (courtesy D. Judge [147]).

pulses. A few microseconds prior to the discharge, the transistor is shut off by a positive square wave pulse of a few volts. The duration of this square wave is sufficiently long to cover the period of the discharge. With the transistor in its nonconducting stage, the light source pulses are detected by the microammeter. The value of the input capacitor of the microammeter can be adjusted for any desired time constant.

Conclusions

To produce radiation below 1000 Å, the most useful source for many applications is the condensed spark discharge in a low pressure gas. Where the pulsed nature of the source cannot be tolerated, the duoplasmatron is the only suitable dc source which can produce many lines of good intensity between 400 and 1000 Å. At shorter wavelengths the only continuous source of line radiation is the characteristic x-radiation produced by bombarding a suitable target with electrons. For example, a carbon target produces the carbon K line at 44 Å, aluminum produces the 8.3 Å Al K-line, while a copper target produces the Cu L-line at 13.3 Å. Above 1000 Å, the molecular spectrum of hydrogen can be used for most applications. The method of excitation is not too important and is best dictated by the availability of equipment; however, the cold cathode glow discharge provides a very stable source and is one that requires practically no maintenance. For the production of special line radiation, the appropriate source and excitation conditions should be selected as described on pp. 129 to 178.

The spectra shown in Sect. 5.3 were all taken with a $\frac{1}{2}$ m Seya monochromator with a 1200 line per mm grating blazed for 600 Å. The band-pass was approximately 2.5 Å. The detector was an EMI 9514S photo-multiplier sensitized with sodium salicylate and calibrated using a rare gas ion chamber. The line intensities, measured in units of photons per sec, refer to the intensity of the radiation emerging from the exit slit of the monochromator. The intensity of the lines produced by a given light source can be directly compared with similar lines produced by a different source. However, the relative intensities of widely separated lines may not represent the actual energy distribution emerging from the light source, since grating efficiencies can vary rapidly as a function of wavelength in the vacuum uv region.

References 149 to 152 provide additional reviews of vacuum uv light sources. Tables of vacuum uv wavelengths have been published by Kelly [153] and Moore [154]. A photographic atlas of atomic spectra between 1100 and 2250 Å has been prepared by Junkes et al. [85] for the elements Al, C, Cu, Fe, Ge, Hg, Si, and H_2. Additional wavelengths of highly ionized Ne, Ar, Kr, and Xe produced by a condensed spark discharge have recently been published by Schönheit [147].

REFERENCES

[1] A. S. Coolidge, H. M. James, and R. D. Present, *J. Chem. Phys.* **4**, 193 (1936).

[2] H. M. James and A. S. Coolidge, *Phys. Rev.* **55**, 184 (1939).

[3] A. S. Coolidge, *Phys. Rev.* **65**, 236 (1944).

[4] J. J. Hopfield, *Phys. Rev.* **35**, 1133 (1930).

[5] J. J. Hopfield, *Phys. Rev.* **36**, 784 (1930).

[6] J. J. Hopfield, *Astrophys. J.* **72**, 133 (1930).

[7] Y. Tanaka, *Sci. Papers Inst. Phys. Chem. Research* (Tokyo) **39**, 465 (1942).

[8] Y. Tanaka, A. S. Jursa, and F. J. LeBlanc, *J. Opt. Soc. Am.* **48**, 304 (1958).

[9] R. E. Huffman, Y. Tanaka, and J. C. Larrabee, Appl. *Optics* **2**, 617 (1963); and *J. Appl. Phys.* (Japan) **4**, Supplement I, 494 (1965).

[10] R. E. Huffman, J. C. Larrabee, Y. Tanaka, and D. Chambers, *J. Opt. Soc. Am.* **55**, 101 (1965).

[11] R. E. Huffman, J. C. Larrabee, and D. Chambers, *Appl. Optics* **4**, 1145 (1965).

[12] R. E. Huffman, W. W. Hunt, Y. Tanaka, and R. L. Novack, *J. Opt. Soc. Am.* **51**, 693 (1961).

[13] R. E. Huffman, Y. Tanaka, and J. C. Larrabee, *J. Opt. Soc. Am.* **52**, 851 (1962).

[14] R. E. Huffman, J. C. Larrabee, and Y. Tanaka, *Appl. Optics* **4**, 1581 (1965).

[15] E. C. Beaty and P. L. Patterson, *Phys. Rev.* **137**, A346 (1965).

[16] A. L. Smith and J. W. Meriwether, *J. Chem. Phys.* **42**, 2984 (1965).

[17] R. E. Huffman, W. W. Hunt, Y. Tanaka, and R. L. Novack, *J. Opt. Soc. Am.* **51**, 693 (1961).

[18] R. E. Huffman, Y. Tanaka, and J. C. Larrabee, *J. Opt. Soc. Am.* **52**, 851 (1962).

[19] T. Takamine, T. Suga, Y. Tanaka, and G. Imotani, *Sci. Papers Inst. Phys. Chem. Research* (Tokyo) **35**, 447 (1939).

[20] Y. Tanaka, A. S. Jursa, and F. J. LeBlanc, *J. Opt. Soc. Am.* **47**, 105 (1957).

[21] Y. Tanaka and M. Zelikoff, *Phys. Rev.* **93**, 933 (1954).

[22] Y. Tanaka and M. Zelikoff, *J. Opt. Soc. Anm.* **44**, 254 (1954).

[23] Y. Tanaka, *J. Opt. Soc. Am.* **45**, 710 (1955).

[24] P. G. Wilkinson, *J. Opt. Soc. Am.* **45**, 1044 (1955).

[25] P. G. Wilkinson and Y. Tanaka, *J. Opt. Soc. Am.* **45**, 344 (1955).

[26] P. G. Wilkinson and E. T. Byram, *Appl. Optics* **4**, 581 (1965).

[27] T. Lyman, *Astrophys. J.* **60**, 1 (1924); *Science* **64**, 89 (1926).

[28] R. E. Worley, *Rev. Sci. Instr.* **13**, 67 (1942).

[29] W. R. S. Garton, *J. Sci. Instr.* **36**, 11 (1959) and **30**, 119 (1953).

[30] W. R. S. Garton, I. W. Celnick, H. Hessberg, and J. E. G. Wheaton, "Proc. 4th Intern. Conf. on Ioniz. Phenomena in Gases" (North-Holland, Amsterdam, 1960), p. 518.

[31] J. E. G. Wheaton, *Appl. Optics* **3**, 1247 (1964).

[32] M. Morlais and S. Robin, *Compt. rend.* **258**, 862 (1964).

[33] R. Goldstein and F. N. Mastrup, *J. Opt. Soc. Am.* **56**, 765 (1966).

[34] A. Lienard, *L'Eclairage Elec.* **16**, 5 (1898).

[35] G. A. Schott, *Electromagnetic Radiation* (Cambridge U.P., Cambridge, 1912).

[36] D. Iwanenko and I. Pomeranchuk, *Phys. Rev.* **65**, 343 (1944).

[37] E. M. McMillan, *Phys. Rev.* **68**, 144 (1945).

[38] J. P. Blewett, *Phys. Rev.* **69**, 87 (1946).

[39] L. I. Schiff, *Rev. Sci. Instr.* **17**, 6 (1946).

[40] J. Schwinger, *Phys. Rev.* **70**, 798 (1946).

[41] J. Schwinger, *Phys. Rev.* **75**, 1912 (1949).

[42] F. R. Elder, A. M. Gurewitsch, R. V. Langmuir, and H. C. Pollock, *Phys. Rev.* **71**, 829 (1947).

[43] F. R. Elder, R. V. Langmuir, and H. C. Pollock, *Phys. Rev.* **74**, 52 (1948).

[44] P. L. Hartman and D. H. Tomboulian, *Phys. Rev.* **91**, 1577 (1953).

[45] D. H. Tomboulian and P. L. Hartman, *Phys. Rev.* **102**, 1423 (1956).

[46] D. H. Tomboulian and D. E. Bedo, *J. Appl. Phys.* **29**, 804 (1958).

[47] K. Codling and R. P. Madden, *J. Appl. Phys.* **36**, 380 (1965).

[48] Y. Cauchois, C. Bonelle, and G. Missoni, *Compt. Rend.* **257**, 409 and 1242 (1963).

[49] W. K. H. Panofsky and M. Phillips, *Classical Electricity and Magnetism* (Addison-Wesley, Cambridge, 1955) pp. 301–307, 2nd ed.

[50] P. Joos, *Phys. Rev. Letters* **4**, 558 (1960).

[51] E. Jahnke and F. Emde, *Tables of Functions* (Dover, New York, 1945).

[52] R. P. Madden and K. Codling, *Phys. Rev. Letters* **10**, 516 (1963).

[53] K. Codling and R. P. Madden, *Phys. Rev. Letters* **12**, 106 (1964).

[54] R. P. Madden and K. Codling, *J. Opt. Soc. Am.* **54**, 268 (1964).

[55] K. Codling and R. P. Madden, *Appl. Optics* **4**, 1431 (1965).

[56] R. P. Madden and K. Codling, *Astrophys. J.* **141**, 364 (1965).

[57] S. T. Stephenson, *Handbuch der Physik*, Vol. 30 (Springer-Verlag, Berlin, 1957) p. 337.

[58] S. T. Stephenson and F. D. Mason, *Phys. Rev.* **75**, 1711 (1949).

[59] H. Neff, *Z. Physik* **131**, 1 (1951).

[60] T. J. Peterson, Jr. and D. H. Tomboulian, *Phys. Rev.* **125**, 235 (1962).

[61] T. J. Peterson, Jr., Thesis, "The Soft X-Ray Continuous Spectrum from Low Energy Electrons in the 80–180 Å Region" (University Microfilms, Ann Arbor, 1961).

[62] A. P. Lukirskii and V. A. Fomichev, *Optics and Spectroscopy* **19,** 441 (1965).

[63] B. L. Henke, *X-Ray Optics and X-Ray Microanalysis*, eds. H. H. Pattee, V. E. Cosslett, and A. Engström (Academic, New York, 1963), p. 157; *Advances X-Ray Analysis* **4,** 244 (1961).

[64] A. J. Caruso and W. M. Neupert, *Rev. Sci. Instr.* **36,** 554 (1965).

[65] A. E. Sandström, *Handbuch der Physik*, Vol. 30 (Springer-Verlag, Berlin, 1957) p. 78.

[66] D. H. Tomboulian, *Handbuch der Physik*, ed. S. Flügge, Vol. 30 (Springer-Verlag, Berlin, 1957) p. 246.

[67] A. W. Ehler and G. L. Weissler, *Appl. Phys. Letters* **8,** 89 (1966).

[68] S. Glasstone and R. Lovberg, *Controlled Thermonuclear Reactions* (Van Nostrand, New York, 1960) pp. 31–33.

[69] S. J. Smith and E. M. Purcell, *Phys. Rev.* **92,** 1069 (1953).

[70] I. M. Frank, *Izv. Akad. Nauk. SSSR, Ser. Fiz.* **6,** 3 (1942).

[71] W. W. Salisbury, U.S. Patent No. 2, 634, 372 (1953).

[72] B. M. Bolotovskii and A. K. Burtsev, *Opt. Spect. (USSR)* **14,** 263 (1965).

[73] K. Ishiguro and T. Tako, *Opt. Acta* **8,** 25 (1961).

[74] C. W. Barnes and K. G. Dedrick, *J. Appl. Phys.* **37,** 411 (1966).

[75] G. Balloffet, *Am. Phys.* **5,** 1256 (1960).

[76] G. Balloffet, J. Romand, and B. Vodar, *Compt. Rend.* **252,** 4139 (1961).

[77] G. Balloffet, J. Romand, and J. Kieffer, *Spectrochimica Acta* **18,** 791 (1962).

[78] B. Lotte, M. Bon, and J. Romand, *J. de Phys. Rad.* **24,** 346 (1963).

[79] J. Romand, *J. Appl. Phys.* (Japan) **4,** Suppl. 1, 506 (1965).

[80] H. Damany, J-Y Roncin, and N. Damany-Astoin, *Appl. Optics* **5,** 297 (1966).

[81] K. T. Compton, *A.I.E.E. Trans.* **46,** 868 (1927).

[82] J. D. Cobine, *Gaseous Conductors* (Dover, New York, 1958), p. 217.

[83] W. R. Hunter, *Proc. 10th Colloquium Spectrosc. Intern.*, 247 (1962).

[84] K. E. Schubert and R. D. Hudson, A Photoelectric Atlas of the Intense Lines of the Hydrogen Molecular Emission Spectrum from 1025 to 1650 Å at a Resolution of 0.10 Å (Report Number ATN-64(9233)-2, Aerospace Corp., Los Angeles, 1963).

[85] J. Junkes, E. W. Salpeter, and G. Milazzo, *Atomic Spectra in the Vacuum Ultraviolet from 2250 to 1100 Å*, Pt. One, Al, C, Cu, Fe, Ge, Hg, Si, and H_2 (Specola Vaticano, Città del Vaticano, 1965).

[86] G. Herzberg and L. L. Howe, *Can. J. Phys.* **37,** 646 (1959).

[87] B. Rosen, *Atlas des longeurs d'onde caracteristiques des bandes d'emission et d'absorption des molecules diatomiques* (Hermann, Paris, 1952) p. 369.

[88] R. K. Curran, *J. Chem. Phys.* **38,** 2974 (1963).

[89] M. P. Teter, F. E. Niles, and W. W. Robertson, *Bull. Am. Phys. Soc.* **11,** 503 (1966).

[90] F. Paschen, *Ann. d. Physik* **50,** 901 (1916).

[91] H. Schuler, *Physik. Zeitschr.* **22,** 264 (1921).

[92] R. G. Newburgh, L. Heroux, and H. E. Hinteregger, *Appl. Optics* **1,** 733 (1962); and R. G. Newburgh, *Appl. Optics* **2,** 864 (1963).

[93] R. D. Deslattes, T. J. Peterson, Jr., and D. H. Tomboulian, *J. Opt. Soc. Am.* **53,** 302 (1963).

[94] W. J. Scouler and E. D. Mills, *Rev. Sci. Instr.* **35,** 489 (1964).

[95] P. Johnson, *J. Opt. Soc. Am.* **42**, 278 (1952).

[96] P. L. Hartman and J. R. Nelson, *J. Opt. Soc. Am.* **47**, 646 (1957).

[97] P. L. Hartman, *J. Opt. Soc. Am.* **51**, 113 (1961).

[98] B. Brehm and H. Siegert, *Z. angew. Phys.* **19**, 244 (1965).

[99] M. V. Ardenne, *Tabellen der Elektronenphysik, Ionenphysik and Ubermikroskopie* (Deutschen Verl. der Wissenschafter, Berlin, 1956) Vol. I, p. 544.

[100] C. D. Moak, H. E. Banta, J. N. Thurston, J. W. Johnson, and R. F. King, *Rev. Sci. Instr.* **30**, 694 (1959).

[101] H. Froehlich, *Nucleonik* **1**, 183 (1959).

[102] B. S. Burton, Jr., *The Duoplasmatron* presented at ARS Conference on Electrostatic Propulsion (November 1960).

[103] C. M. Braams, P. Zieske, and M. J. Kofoid, *Rev. Sci. Instr.* **36**, 1411 (1965).

[104] J. A. R. Samson and H. Liebl, *Rev. Sci. Instr.* **33**, 1340 (1962).

[105] L. Minnhagen, *J. Res.* NBS **68C**, 237 (1964); *see also* L. Minnhagen, B. Petersson, and L. Stigmark, *Arkiv. Fysik* **16**, 541 (1960).

[106] M. Plato, *Z. Naturforsch* **19a**, 1324 (1964).

[107] W. R. S. Garton, M. S. W. Webb, and P. C. Wildy, *J. Sci. Instr.* **34**, 496 (1957).

[108] G. Dieke and S. Cunningham, *J. Opt. Soc. Am.* **42**, 187 (1952).

[109] M. Zelikoff, P. Wyckoff, L. M. Aschenbrand, and R. Loomis, *J. Opt. Soc. Am.* **42**, 818 (1952).

[110] P. G. Wilkinson and Y. Tanaka, *J. Opt. Soc. Am.* **45**, 344 (1955).

[111] P. G. Wilkinson, *J. Opt. Soc. Am.* **45**, 1044 (1955).

[112] N. N. Axelrod, *J. Opt. Soc. Am.* **53**, 297 (1963).

[113] E. W. Schlag and F. J. Comes, *J. Opt. Soc. Am.* **50**, 866 (1960).

[114] P. Warneck, *Appl. Optics* **1**, 721 (1962).

[115] H. Okabe, *J. Opt. Soc. Am.* **54**, 478 (1964).

[116] L. J. Stief and R. J. Mataloni, *Appl. Optics* **4**, 1674 (1965).

[117] F. C. Fehsenfeld, K. M. Evenson, and H. P. Broida, *Rev. Sci. Instr.* **36**, 294 (1965); Nat. Bur. Std. Report No. 8701 (November 1964).

[118] R. A. Millikan and R. A. Sawyer, *Phys. Rev.* **12**, 167 (1918).

[119] T. Lyman, *The Spectroscopy of the Extreme Ultraviolet* (Longmans, Green New York, 2nd ed. 1928) p. 46.

[120] B. Vodar and N. Astoin, *Nature* **166**, 1029 (1950).

[121] K. Bockasten, *Arkiv Fysik* **9**, 457 (1955).

[122] N. Astoin, J. Romand, and B. Vodar, *The Threshold of Space*, ed. M. Zelikoff (Pergamon, New York, 1957) p. 128.

[123] N. Astoin, *Compt. rend.* **234**, 2050 (1952).

[124] J. Romand and G. Balloffet, *J. phys. radium* **16**, 489 (1955); *Compt. rend.* **244**, 739 (1957).

[125] G. Balloffet and J. Romand, *J. phys. radium* **16**, 490 (1955).

[126] B. Vodar and J. Romand, *Mikrochim. Acta* **32**, 429 (1955).

[127] J. Romand and B. Vodar, *Spectrochim Acta.* **8**, 229 (1956).

[128] G. Balloffet, *Ann. phys.* **5**, 1243 (1960).

[129] G. Balloffet and I. Manescu, *J. phys.* **25**, 701 (1964).

[130] G. Balloffet, *J. phys.* **25**, 73A (1964).

[131] G. H. C. Freeman, *Proc. Phys. Soc.* **86**, 117 (1965).

[132] R. C. Gibbs, A. M. Vieweg, C. W. Gartlein, *Phys. Rev.* **34**, 406 (1929).

[133] B. C. Fawcett, B. B. Jones, and R. Wilson, *Proc. Phys. Soc.* **78**, 1223 (1961).

[134] B. C. Fawcett, A. H. Gabriel, B. B. Jones, and N. J. Peacock, *Proc. Phys. Soc.* **84**, 257 (1964).

[135] H. J. Karr, E. A. Knapp, and J. E. Osher, *Phys. Fluids* **4**, 424 (1961).

[136] V. D. Pis'mennyi, I. M. Podgornyi, and Sh. Sukever, *Soviet Physics* JETP **16**, 1416 (1963).

[137] R. Wilson, *Ann. Astrophys.* **27**, 771 (1964).

[138] F. W. Paul, *Phys. Rev.* **56**, 1067 (1939).

[139] Po Lee and G. L. Weissler, *J. Opt. Soc. Am.* **42**, 80 (1952).

[140] N. Wainfan, W. C. Walker, and G. L. Weissler, *J. Appl. Phys.* **24**, 1318 (1953).

[141] R. Ermisch and E. Schönheit, *Z. Naturforsch.* **20a**, 611 (1965).

[142] O. P. Rustgi, *J. Opt. Soc. Am.* **55**, 630 (1965).

[143] G. L. Weissler, *J. Appl. Phys.* (Japan) **4**, Supplement I, 486 (1965).

[144] Po Lee, *J. Opt. Soc. Am.* **55**, 783 (1965).

[145] J. A. R. Samson and R. B. Cairns, *J. Geophys. Res.* **69**, 4583 (1964).

[146] J. A. R. Samson, *Advances in Atomic and Molecular Physics*, ed. D. Bates and I. Estermann (Academic, New York, 1966), p. 177.

[147] E. Schönheit, *Optik* **23**, 409 (1965/1966).

[148] D. Judge, Fluorescence Spectra of the Molecular Ion $O_2{}^+$, $N_2{}^+$, and CO^+ Excited by Vacuum Ultraviolet Radiation (Thesis, University of Southern California, Los Angeles, 1965).

[149] F. J. Comes, *Z. Instrumentenk.* **68**, 69 (1960).

[150] J. C. Boyce, *Rev. Mod. Phys.* **13**, 1 (1941).

[151] J. A. R. Samson, *Vacuum Ultraviolet Light Sources*, NASA CR-17 (Off. of Tech. Serv., Washington, 1963).

[152] H. E. Blackwell, G. S. Shipp, M. Ogawa, and G. L. Weissler, *J. Opt. Soc. Am.* **56**, 665 (1966).

[153] R. L. Kelly, *Vacuum Ultraviolet Emission Lines*, (UCRL 5612, Univ. Calif. Lawrence Rad. Lab. 1960).

[154] C. E. Moore, *Atomic Energy Levels* (Nat. Bur. Stand. Circular 467, Vols. I, II, and III, 1949–1958); also *An Ultraviolet Multiplet Table* (Nat. Bur. Stand. Circular 488, 1950–1962).

6

Filter and Window Materials

The discussion of the transparency of materials to vacuum uv radiation is divided into three parts. The first part deals with materials which are rugged and are normally used as windows capable of withstanding a pressure of an atmosphere or more. The second part deals with materials more suitably described as filters. These are thin metallic or plastic films which are quite fragile when unsupported and generally cannot withstand differential pressures of more than a few torr. The third part discusses the use of gas filters.

6.1 CRYSTALLINE MATERIALS

The most commonly used window material in the vacuum uv region is lithium fluoride (LiF). It has the shortest wavelength transmission limit (1040 Å) of any known window material with the exception of the thin films discussed below. Because of its good transmission properties between 1040 and 2000 Å, it is often used to the exclusion of all other materials. The transparency of LiF was first discovered by Melvin in 1931 [1]. He was guided in his selection of LiF by the fact that the index of refraction of a material is closely related to its absorption properties. He observed that if the dispersion curves of LiF, CaF_2, and the alkali halides were plotted as a function of wavelength, none of the curves crossed and, all were of the same general shape. Thus he chose NaF and LiF for his transmission measurements since they had the lowest indices of refraction of any cubic crystal measured at 5893 Å, namely, 1.328 for NaF and 1.384 for LiF. His results gave a transmission limit of 1083 Å for his best LiF crystal 1 mm thick. The NaF crystal (also 1 mm thick) had a transmission limit of 1320 Å. On the basis of its refractive index, Melvin felt that NaF should be more transparent than LiF and suggested that impurities within the crystal were at fault. However, if we correlate the molecular weight and density of a fluoride crystal with its short wavelength transmission limit as given below, we see that as the molecular weight and density decrease, so does the wavelength of the transmission limit. Since the molecular weight of NaF is greater than that of LiF, its transmission limit might also be expected to lie at longer wavelengths. Other suitable window materials, with their short wavelength transmission limits quoted in parentheses, are MgF_2 (1120 Å), CaF_2 (1220 Å), SrF_2 (1280 Å), BaF_2 (1340 Å), sapphire (1410 Å), and synthetic fused quartz (1600 Å). The transmittance of all of these windows (with the exception of MgF_2) has

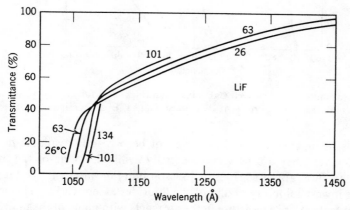

Fig. 6.1 Transmittance of cleaved LiF, 1.95 mm thick, as a function of wavelength for various temperatures (courtesy A. H. Laufer et al. [4]).

been found to be temperature dependent [2–5]. In all probability, the transmittance of MgF_2 will also be found to be temperature dependent. However, this parameter has not yet been measured. As the temperature increases over the range −196°C to +380°C, the transmission limit shifts to longer wavelengths at a rate of approximately $\frac{1}{3}$ to $\frac{1}{4}$ Å per degree centigrade. Curves showing the transmittance variation with the temperature for LiF, CaF_2, BaF_2, and sapphire are shown in Figs. 6.1 to 6.4.

The transmittance of a 1 mm thick crystal of MgF_2 is shown in Fig. 6.5. The percent transmittance at 1120 Å is 0.4 per cent. However, this value may vary depending on the purity of the crystal. Duncanson and Stevenson [6] quote a value of 7 per cent at 1100 Å for a crystal 1.9 mm thick. Johnson [7] has measured the transmittance of a MgF_2 polarizer 3 mm thick and found a maximum value of about 37 per cent for the E and O rays, separately, at wavelengths greater than 1500 Å. He quotes a transmission cutoff at about 1125 Å (see Chapter 9). Thin evaporated films of magnesium fluoride have been studied by Fabre and Romand [8,9], who have measured the values of the optical constants between 950 and 1600 Å.

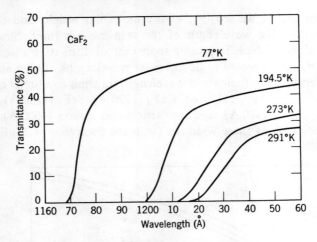

Fig. 6.2 Transmittance of cleaved CaF_2, 0.71 mm thick, as a function of wavelength for various temperatures (courtesy A. R. Knudson and J. E. Kupperian [2]).

Crystals of magnesium fluoride have not been readily available in the past. However, they are now commercially available. One of the most attractive features about MgF_2 is its low solubility in water. In fact, it has about the lowest solubility of any of the fluorides except CaF_2. This is an important feature since moisture in the air reacts with many of the fluorides reducing their transmission. Studies on evaporated films of MgF_2 by Hunter [10] and by Canfield et al. [11] have shown very little aging of the films when they were exposed to air for prolonged periods.

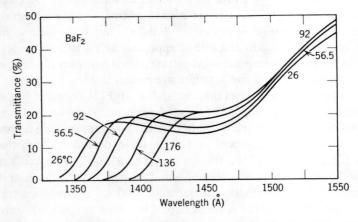

Fig. 6.3 Transmittance of cleaved BaF_2, 1.95 mm thick, as a function of wavelength for various temperatures (courtesy A. H. Laufer et al. [4]).

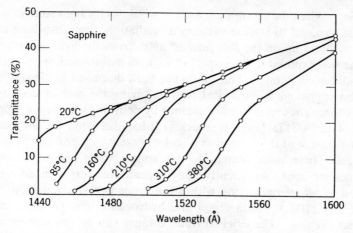

Fig. 6.4 Transmittance of synthetic sapphire, 0.8 mm thick, as a function of wavelength for various temperatures (courtesy S. A. Yakovlev [3]).

The transmission properties of LiF deteriorate when the crystal is exposed to moist air for lengthy periods of time. The transmission further deteriorates when the crystal is exposed to intense sources of vacuum uv radiation. Patterson and Vaughan [12] have shown that unless LiF is stored in an exceptionally dry area, its transmittance will decrease. They showed that moisture reacted with the crystal probably forming hydrofluoric acid (HF) and other products, thereby reducing the transmittance. Heating the crystals in an oven to about 500°C generally restores the original transmittance. Davis [5] has shown that cleaning the crystals with ethyl alcohol in an ultrasonic cleaner can also restore the transmission

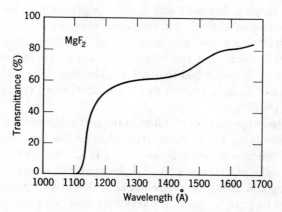

Fig. 6.5 Transmittance of polished MgF_2, 1 mm thick, as a function of wavelength.

properties. When the crystals are used as windows for light sources or are otherwise exposed to intense vacuum uv radiation, thin absorbing layers frequently build up on the side *farthest* away from the light source; that is, on the side exposed to the imperfect vacuum maintained by mechanical and oil-diffusion pumps. This effect has been discussed by Taylor et al. [13], who suggested an empirical relation giving the rate of decrease in transmission. The crystals can be cleaned by polishing with Linde sapphire powder, type A-5175 [14]. Warneck [15] has discussed the decrease in transmission due to the formation of color centers by argon and hydrogen discharges. These color centers do not appear when xenon or krypton discharges are used. Apparently, they appear when LiF is irradiated by radiation of wavelength lying within the fundamental absorption band, namely, 900–1100 Å. Both argon and hydrogen discharges emit in this wavelength region. The original transmittance can be restored again by annealing the crystals at 500°C.

Heath and Sacher [15a] have investigated the effect on the transmittance of LiF, CaF_2, MgF_2, BaF_2, Al_2O_3, and fused SiO_2 after irradiation by 10^{14} electrons per cm^2 of 1 and 2 MeV energy. LiF became almost opaque after irradiation. Most of the other materials suffered some loss in transmittance below 2000 Å. BaF_2, the exception, showed no change in transmittance.

6.2 THIN FILMS

Organic compounds–such as Zapon (cellulose acetate); Collodion, Parlodion, and Celluloid (all belonging to the cellulose nitrate family); Mylar (polyethylene tetraphthalate); Formvar (polyvinyl formal); and other plastics–can make suitable windows and supports for transmittance studies. The cellulose compounds can be dissolved or diluted with amyl acetate, while ethylene bromide will dissolve Mylar. Thin films are made by allowing a drop of the solution to spread over the surface of distilled water. If interference colors can be observed, the films are generally in excess of 5000 Å thick, and the solution should be further thinned. A suitable thickness is about 500 to 1000 Å. The thin films can be picked up with fine mesh screens, preferably the electro-formed type with a high transmittance ($\sim$70%).

The transmission properties of a film in the soft x-ray spectral region are governed primarily by the K absorption edges of the constituents of the film. Most of the plastic materials consist of H, C, N, and O atoms. The K absorption edge of carbon is 43.77 Å, oxygen has an edge at 23.37 Å, while the nitrogen K absorption edge lies at 31.05 Å. The transmittance of a material is at its peak to the long wavelength side of an absorption edge. The energies of the absorption edges of the elements have been tabulated by

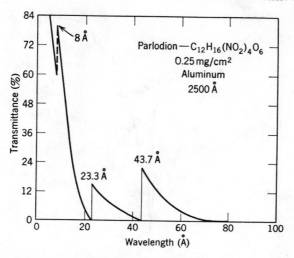

Fig. 6.6 Transmittance of Parlodion coated with an aluminum film 2500 Å thick as a function of wavelength (courtesy T. A. Chubb et al. [21]).

Sandström [16], Tomboulian [17], and Cosslett and Nixon [18]. Mass absorption coefficients have been listed by Victoreen [18] and by Henke et al. [20]. Figure 6.6 shows the computed transmittance of Parlodion coated with an aluminum film 2500 Å thick [21]. The K absorption edges of C, O, and Al can be seen. Hunter et al. [22] and O'Bryan [23] have measured the transmittances of Parlodion and celluloid, respectively. Their results are reproduced in Fig. 6.7, while Fig. 6.8 shows the linear absorption coefficient μ of Zapon as measured by Tomboulian and Bedo [24]. The transmittance T of a material is related to the thickness x and to the absorption coefficient by the relation

$$T = \frac{I}{I_0} = \exp\left(-\mu x\right), \tag{6.1}$$

where I and I_0 are the transmitted and incident intensities, respectively. Table 6.1 expresses the value in per cent of T for various values of μx.

Table 6.1 Transmittance T of a Material of Thickness x and Linear Absorption Coefficient μ as Determined from the Relation $T = I/I_0 = \exp\left(-\mu x\right)$

μx	0.1	0.2	0.3	0.4	0.5	0.6	0.7	0.8	0.9	1.0	2.0	3.0	4.0	4.6	5.0
T (%)	90	82	74	67	61	55	50	45	41	37	14	5	1.8	1.0	0.7

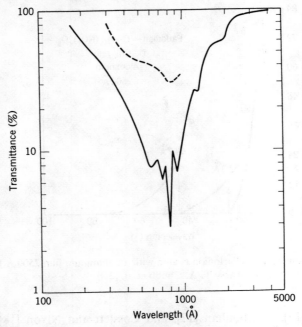

Fig. 6.7 Transmittance of a Parlodion film approximately 270 Å thick, solid curve; and of a celluloid film 100 Å thick as measured by O'Bryan [23], dashed curve (courtesy W. R. Hunter et al. [22]).

Thin metallic films provide the major source of filters for the region below 1040 Å. The first measurements in the vacuum uv of the transmittance of a metal film were made by Tomboulian and Pell [25] in 1951. By evaporating aluminum onto a Zapon support, they measured its transmittance between 80 and 320 Å. Astoin and Vodar [26], using Collodion as a support, extended the measurements to 750 Å.

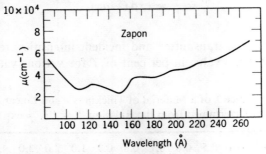

Fig. 6.8 Absorption coefficient of Zapon as a function of wavelength (courtesy D. H. Tomboulian and D. E. Bedo [24]).

The free-electron model of a metal actually predicts that at some critical frequency ν_p the metal will change from a reflecting to a transmitting medium. The older theories picture the electrons interacting *individually* with the incident radiation and neglect long-ranged Coulomb interactions between other electrons, whereas the more recent theory of Bohm and Pines [27] considers the electrons interacting collectively with the incident radiation. According to the collective model, the electron plasma is able to support an electromagnetic wave at frequencies greater than ν_p. Both theories, however, give for the plasma frequency

$$\nu_p = \left(\frac{ne^2}{\pi m}\right)^{\frac{1}{2}}$$
$$= (9 \times 10^3)n^{\frac{1}{2}}, \tag{6.2}$$

where m and e denote the electronic mass and charge, respectively; while n is the number of electrons per cm^3 that take part in the collective oscillations and is generally taken to be the density of the valence electrons. However, in some atoms, the inner shell electrons are not bound so tightly as they are in other atoms and can share in the oscillations. Thus, the appropriate value of n is not always predictable; however, in many cases, the onset of transmission as given by (6.2) agrees with the experimental results. Where no transmission data exist, (6.2) can be used as a guide. Table 6.2 lists the plasma frequency as calculated from (6.2) for a number of elements. The plasma frequencies for the noble metals gold, silver, and copper, along with zinc and cadmium, have been determined twice by

Table 6.2 Calculated Transmission Onsets for Various Elements

Element	Be	B	C	Mg	Al	Si	Ti	Cr	Mn	Fe	Co
z	2	3	4	2	3	4	4	6	7	8	9
$h\nu_p$ (eV)	19	24	25	11	16	17	18	25	28	31	34
(Å)	653	517	496	1127	775	729	689	496	443	400	365

Element	Ni	Ge	Mo	In	Sn	Sb	Te	Ta	Pt	Pb	Bi
z	10	4	6	3	4	5	6	5	10	4	5
$h\nu_p$ (eV)	35	16	23	13	14	15	16	20	30	13	14
(Å)	354	775	539	954	886	827	775	620	413	954	886

Element	Cu	Cu	Zn	Zn	Ag	Ag	Cd	Cd	Au	Au
z	1	11	2	12	1	11	2	12	1	11
$h\nu_p$ (eV)	11	36	13	32	9	30	11	28	9	30
(Å)	1127	344	954	387	1378	413	1127	443	1378	413

considering (a) only the *s* electrons in the outermost shell as being operative, and (b) considering these *s* electrons and the ten *d* electrons in the next inner sub-shell to be involved in the plasma oscillations. Transmission onsets for these materials have not been found at those predicted frequencies. A more detailed discussion of the interaction between the free electrons and the incident radiation can be found in review articles by D. Pines [28].

During the last few years data have been accumulating on the transmissitivity of thin metal films in the vacuum uv. [22,29–46]. The transmittance and absorption coefficients of all of the elements studied so far are given in Figs. 6.9 to 6.37. Rustgi [42] has shown that self supporting films of silver and gold 300 and 1150 Å thick, respectively, are opaque to radiation down to 250 Å. This may be accounted for by the fact that the $N_{IV,V}$ absorption edge of silver (6 eV) and the $O_{IV,V}$ edge of gold (4 eV) precedes the energy for the expected plasma oscillations.

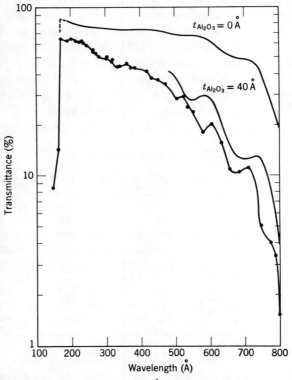

Fig. 6.9 Transmittance of aluminum 800 Å thick. Included for comparison are curves for aluminum calculated from the optical constants, for the case of no oxide, and for 40 Å oxide on both surfaces (courtesy W. R. Hunter et al. [22]).

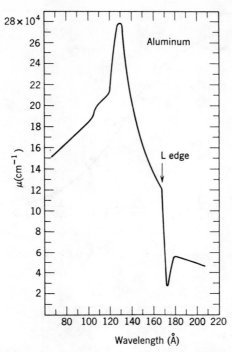

Fig. 6.10 Absorption coefficient of aluminum as a function of wavelength (courtesy D. H. Tomboulian and D. E. Bedo [24]).

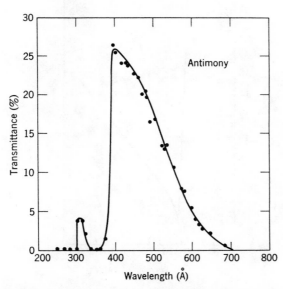

Fig. 6.11 Transmittance of antimony 1000 ± 100 Å thick as a function of wavelength (courtesy O. P. Rustgi [42]).

189

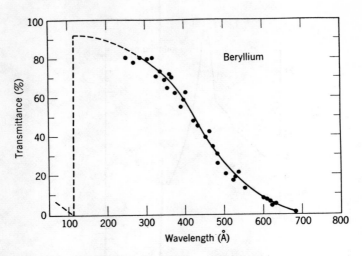

Fig. 6.12 Transmittance of beryllium 875 ± 100 Å thick as a function of wavelength (courtesy O. P. Rustgi [42]).

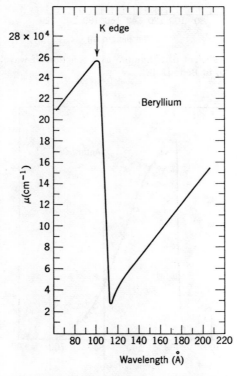

Fig. 6.13 Absorption coefficient of beryllium as a function of wavelength (courtesy D. H. Tomboulian and D. E. Bedo [24]).

190

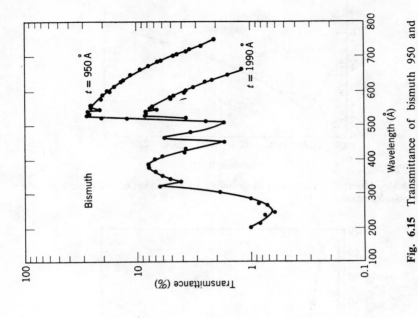

Fig. 6.15 Transmittance of bismuth 950 and 1900 Å thick (courtesy W. R. Hunter et al. [22]).

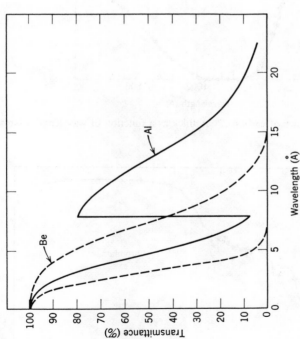

Fig. 6.14 Calculated transmittance of Al and Be between 0 and 20 Å. The solid curve refers to an Al film 2.5 microns thick, while the two dashed curves refer to Be films 25 microns and 260 microns thick. Below the aluminum *K* edge (7.95 Å), the transmittance of a 2.5 micron Al film is identical to a Be film 130 microns thick (courtesy R. L. F. Boyd, *Space Sci. Revs.* **4**, 35 [1965]).

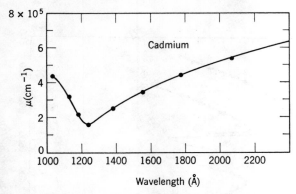

Fig. 6.16 Absorption coefficient of cadmium as a function of wavelength (courtesy S. Robin-Kandare et al. [37]).

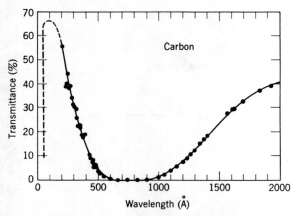

Fig. 6.17 Transmittance of carbon 270 Å thick as a function of wavelength (Samson and Cairns [41]).

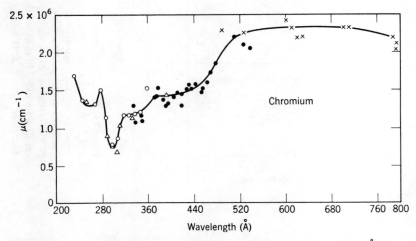

Fig. 6.18 Absorption coefficient of chromium between 200 and 800 Å (courtesy N. N. Axelrod and M. P. Givens [35]).

192

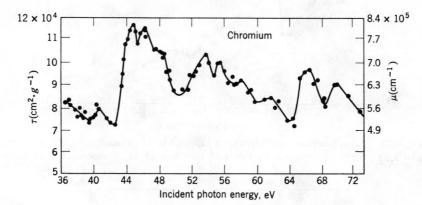

Fig. 6.19 Mass absorption coefficient τ cm^2 . g^{-1} for a chromium film of surface density 27 μg . cm^{-2} measured as a function of the incident photon energy in eV. To obtain the linear absorption coefficient μ, multiply τ by the density $\rho = 7$ g . cm^{-3} (courtesy Tomboulian et al. [32]).

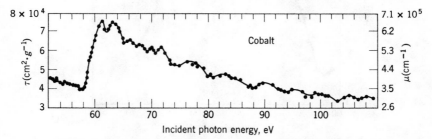

Fig. 6.20 Mass absorption coefficient for a cobalt film of surface density 22 μg . cm^{-2}: $\rho = 8.8$ g . cm^{-3}. See caption to Fig. 6.19 (courtesy D. H. Tomboulian et al. [32]).

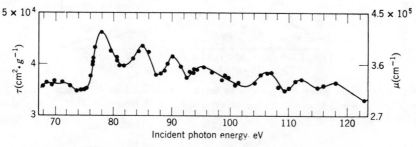

Fig. 6.21 Mass absorption coefficient for a copper film of surface density 88 μg . cm^{-2}: $\rho = 8.9$ g . cm^3. See caption to Fig. 6.19 (courtesy D. H. Tomboulian et al. [32]).

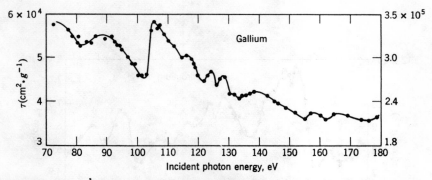

Fig. 6.22 Mass absorption coefficient for a gallium film of surface density 57 μg . cm^{-2}: $\rho = 5.9$ g . cm^{-3}. See caption to Fig. 6.19 (courtesy D. H. Tomboulian et al. [32]).

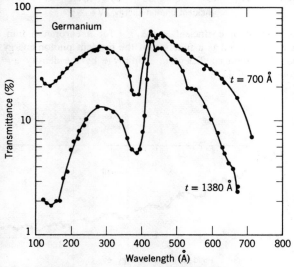

Fig. 6.23 Transmittance of germanium films 700 and 1380 Å thick (courtesy W. R. Hunter et al. [22]).

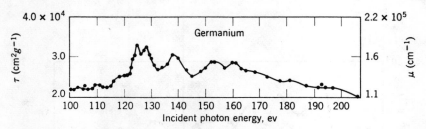

Fig. 6.24 Mass absorption coefficient for a germanium film of surface density 63 μg . cm^{-2}: $\rho = 5.46$ g . cm^{-3}. See caption to Fig. 6.19 (courtesy D. H. Tomboulian et al. [32]).

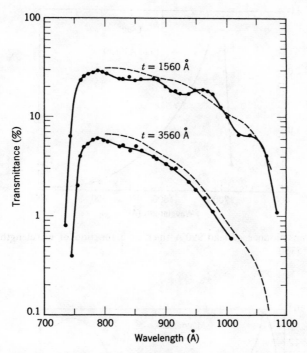

Fig. 6.25 Transmittance of indium films 1560 and 3650 Å thick. Dashed lines are calculated values (courtesy W. R. Hunter et al. [22]).

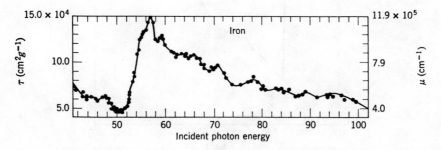

Fig. 6.26 Mass absorption coefficient for an iron film of surface density 19 μg . cm^{-2}: $\rho = 7.9$ g . cm^{-3}. See caption to Fig. 6.19 (courtesy D. H. Tomboulian et al. [32]).

195

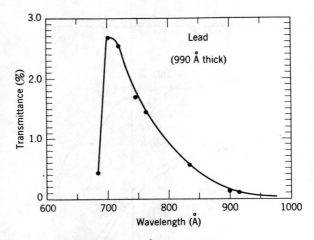

Fig. 6.27 Transmittance of lead 990 Å thick as a function of wavelength (courtesy W. C. Walker [38]).

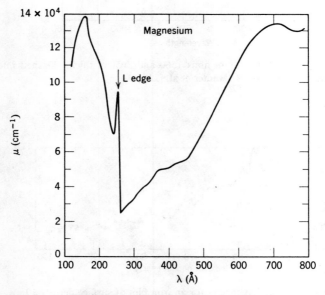

Fig. 6.28 Absorption coefficient of magnesium as a function of wavelength (courtesy D. H. Tomboulian and associates [44, 24]).

196

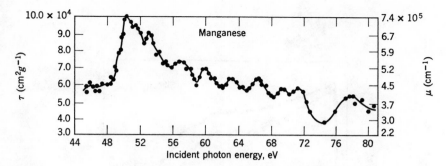

Fig. 6.29 Mass absorption coefficient for a manganese film of surface density 21 μg . cm^{-2}: $\rho = 7.42$ g . cm^{-3}. See caption to Fig. 6.19 (courtesy **D. H. Tomboulian** et al. [32]).

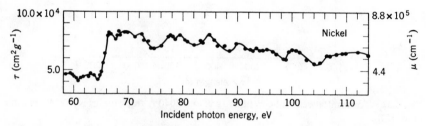

Fig. 6.30 Mass absorption coefficient for a nickel film of surface density 22 μg . cm^{-2}: $\rho = 8.6$ to 8.9 g . cm^{-3}. See caption to Fig. 6.19 (courtesy D. H. Tomboulian et al. [32]).

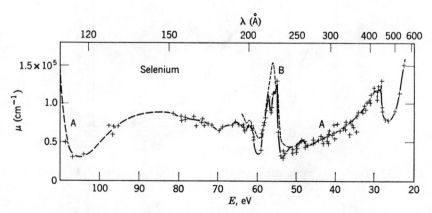

Fig. 6.31 Absorption coefficient of amorphous selenium (courtesy B. Vodar [45]).

197

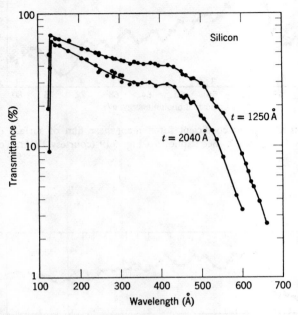

Fig. 6.32 Transmittance of silicon films 1250 and 2040 Å thick (courtesy W. R. Hunter et al. [22]).

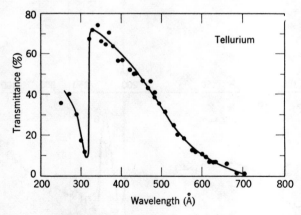

Fig. 6.33 Transmittance of tellurium 550 ± 50 Å thick (courtesy O. P. Rustgi [42]).

198

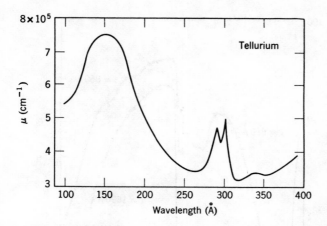

Fig. 6.34 Absorption coefficient of tellurium as a function of wavelength (courtesy R. W. Woodruff and M. P. Givens [30]).

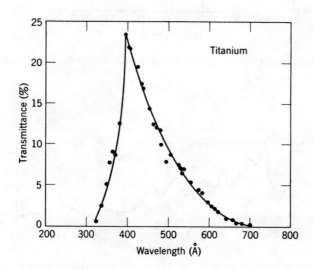

Fig. 6.35 Transmittance of titanium 525 ± 50 Å thick (courtesy O. P. Rustgi [42]).

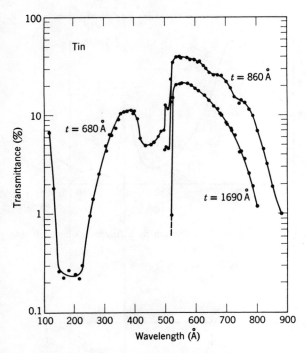

Fig. 6.36 Transmittance of tin films 680, 860, and 1690 Å thick (courtesy W. R. Hunter et al. [22]).

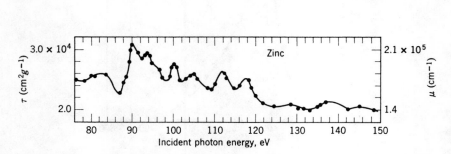

Fig. 6.37 Mass absorption coefficient for a zinc film of surface density 143 μg . cm^{-2}: $\rho = 6.9$ g . cm^{-3}. See caption to Fig. 6.19 (courtesy D. H. Tomboulian et al. [32]).

Thin films of silicon monoxide (SiO) and aluminum oxide (Al_2O_3) have been used as substrates because of their ability to withstand thermal shocks better than thin plastic films. They have also been used successfully as windows, especially in systems to isolate monochromators from experiments requiring clean high vacuum conditions. The transmittance of films about 1000 Å thick is typically only a few per cent. The absorption coefficients for SiO, as measured by Iguchi [47] over the range 600 to 1000 Å, are shown in Fig. 6.38. Astoin and Vodar [48] have also measured the transmittance of SiO; however, they did not measure the film thickness.

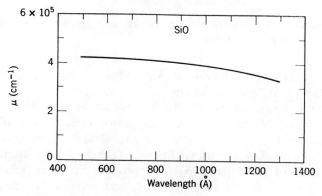

Fig. 6.38 Absorption coefficient of silicon monoxide as a function of wavelength (courtesy Y. Iguchi [47]).

Their results show that the absorption coefficient is relatively constant over the range 140 to 1300 Å. If the results of Bradford et al. [49] can be extrapolated into the vacuum uv region, it would appear that the transparency of SiO films increases when irradiated by intense uv sources. Their results at 2000 Å show a dramatic increase in transmittance from about 10 to 80 per cent after irradiation for two hours in air.

Figure 6.39 shows the absorption coefficient of a 300 Å film of Al_2O_3 over the range 600 to 1000 Å as obtained by Metzger and Watanabe [50]. The production of such thin Al_2O_3 films has been described by several authors [51–54]. The method consists of cutting ordinary household aluminum foil into suitable strips and producing Al_2O_3 by anodization. The aluminum oxidizes by an amount of 13.7 Å per volt in a solution of 22.6 gm of ammonium citrate and 19.2 gm of citric acid in one liter of water. Various anodizing processes have been described by Holland [55]. If one side of the aluminum foil is coated with a paraffined paper made to adhere to the foil by heating, then the coated area will not be anodized [51]. On removing the paper, the exposed aluminum can be dissolved in a

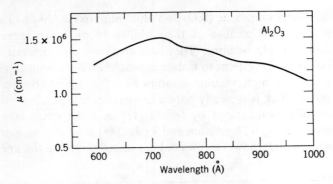

Fig. 6.39 Absorption coefficient of aluminum oxide as a function of wavelength (courtesy P. Metzger and K. Watanabe [50]).

solution of 18 per cent HCl, leaving an Al_2O_3 film, the thickness of which depends on the voltage used in the anodizing process.

The thin films shown in Figs. 6.9 to 6.37 were all prepared by vacuum evaporation. Some of the films were evaporated directly onto plastic films, while others were floated off from their substrates and mounted on electro-formed mesh of known transmittance. Techniques of evaporation have been described by Holland [55]. To produce thin films relatively free from pinholes, it is necessary to remove and prevent dust particles from sticking to the substrate. Even in a clean evaporator it is desirable to keep a close fitting shutter over the substrate during evacuation of the evaporator as a certain amount of dust can be stirred up at that time [56].

Thickness determinations of thin films are most accurately and easily performed using a modified Fizeau-type interferometer as described by Tolansky [57]. Figure 6.40 shows the principle of the interferometer. The thin film, whose thickness is to be measured, can be evaporated onto a microscope slide or an optical flat, a portion of which is covered to prevent any coating with the film. On removing the cover, the entire slide is then coated with a thick layer of silver. Thus a step exists on the silver layer of exactly the same height as the thickness of the underlying film. An optical flat, flat to at least $\frac{1}{10}\lambda$ on one side, is coated on this side with a partially transparent silver layer whose reflectance is approximately 80 per cent. When the flat is placed on top of the film to be studied, as in Fig. 6.40, the multiple reflections that take place in the small air gap between the two surfaces interfere, and a series of fringes appear. The condition for destructive interference is given by

$$m\lambda = 2n_0t \cos \Theta, \tag{6.3}$$

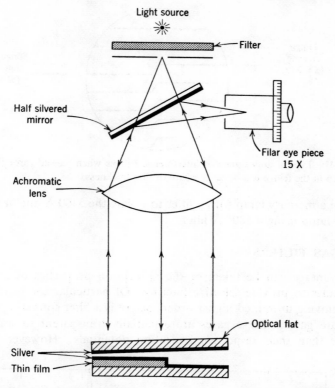

Light source

Filter

Half silvered
mirror

Filar eye piece
15 X

Achromatic
lens

Optical flat

Silver

Thin film

Fig. 6.40 Principle of Tolansky interferometer for measuring the thickness of thin films.

where n_0 is the index of refraction of the air gap and is equal to unity, t is the thickness of the air gap, Θ is the angle of incidence of the mono-chromatic light, and m is an integer. When the light is rendered parallel by the lens and is incident along the normal to the surface of the flat, the thickness of the air gap is given by $t = m(\lambda/2)$. The difference in the thickness of the air gap across the step in the film is, therefore,

$$\Delta t = \Delta m \left(\frac{\lambda}{2} \right), \tag{6.4}$$

where Δt is the thickness of the film, and Δm is the fractional fringe shift given by $\Delta m = s/d$, where s is the fringe shift and d the separation between fringes measured in arbitrary units. Figure 6.41 shows the typical appearance of the fringes as viewed through the eyepiece (usually a filar eyepiece with a magnification of $15\times$). By photographing the fringes, a shift of the order of $\frac{1}{50}$ of a fringe can be measured. The light source

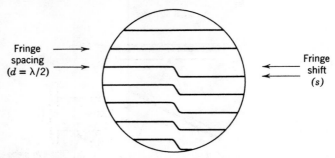

Fig. 6.41 Typical appearance of interference fringes when viewed through eyepiece. The step in the fringe is a direct measure of film thickness.

can be a mercury lamp with a filter to isolate the 5460 Å line or a sodium vapor lamp using a 5890 Å filter.

6.3   GAS FILTERS

Advantage can be taken of the absorption properties of atoms and molecules to provide selective filtering. Of particular use in identifying or removing unwanted higher order spectra is a filter consisting of one of the rare gases. These gases are essentially transparent to wavelengths longer than their respective ionization potentials. However, they are

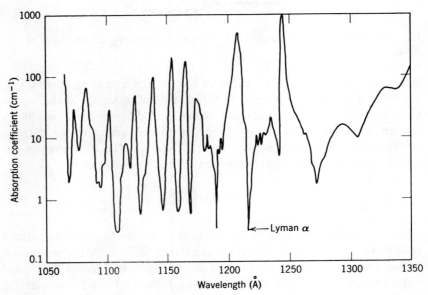

Fig. 6.42 Absorption coefficients of molecular oxygen between 1050 and 1350 Å (courtesy K. Watanabe [58]).

strongly absorbing to shorter wavelengths. Thus with an absorption cell of 20 cm length filled with argon at a pressure ∼500 μ, no radiation between 350 and 787 Å will be transmitted. The spectrum will, therefore, be free from second order lines from 787 to 1574 Å. The change in the spectrum with and without such a filter identifies the higher order lines. The absorption spectra of the rare gases are shown in Chapter 8.

The Lyman-α line of atomic hydrogen (1215.7 Å) happens to fall in a region of extremely low absorption in O_2. Molecular oxygen, in addition, absorbs strongly at neighboring wavelengths (see Fig. 6.42). Thus O_2 provides a particularly useful filter for the Lyman-α line. The absorption

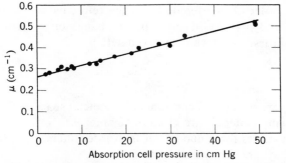

Fig. 6.43 Pressure dependence of the absorption coefficient of O_2 at 1216 Å (courtesy K. Watanabe [58]).

cross section of O_2 at 1215.7 Å is 1×10^{-20} cm². However, the cross section increases with pressure due probably to the absorption of O_4 produced as the oxygen pressure increases. Figure 6.43 shows the variation in cross section with pressure at Lyman-α [58].

Many other gases can be used as filters, but a detailed knowledge of their absorption spectra is necessary.

6.4 INTERFERENCE FILTERS

No narrow band interference filters exist for the vacuum uv. However, Bates and Bradley [59] have shown that it is possible to make broad band-pass filters using alternate layers of evaporated aluminum and magnesium fluoride. They evaporated the layers onto a fused silica substrate (Spectrosil B). The arrangement of the layers is shown in Fig. 6.44. The thickness of each Al layer was about 200 to 300 Å. The optimum thickness d of the MgF_2 spacer for maximum transmission at a wavelength λ_0 was calculated from the relation

$$nd = \left(\frac{\lambda_0}{2\pi}\right)[(m - 1)\pi + \beta], \qquad m = 1, 2, \ldots, \qquad (6.5)$$

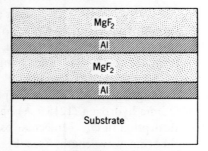

Fig. 6.44 Schematic diagram of MgF$_2$-Al interference filter.

where m is the order of interference, n is the refractive index of the dielectric spacer, and β is the absolute phase change on reflection at the MgF$_2$-Al boundaries. β is given approximately by

$$\tan \beta = \frac{2nk_1}{(n^2 - n_1{}^2 - k_1{}^2)}, \qquad (6.6)$$

where n_1 and k_1 are the optical constants of the metal spacer. The optical constants for many materials are given in Chapter 9.

Transmission curves for a number of first-order filters, as obtained by Bates and Bradley, are shown in Fig. 6.45.

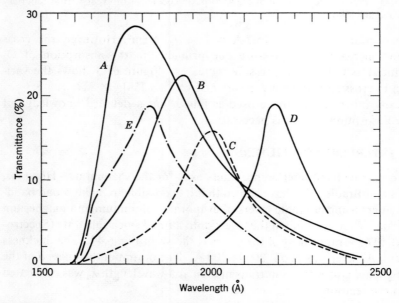

Fig. 6.45 Experimental transmission curves of first-order filters deposited on Spectrosil *B* substrates (courtesy Bates and Bradley [59]).

REFERENCES

[1] E. H. Melvin, *Phys. Rev.* **37**, 1230 (1931).

[2] A. R. Knudson and J. E. Kupperian, *J. Opt. Soc. Am.* **47**, 440 (1957).

[3] S. A. Yakovlev, *Instr. Experimental Tech.* (English trans.) **2**, 396 (1962).

[4] A. H. Laufer, J. A. Pirog, and J. R. McNesby, *J. Opt. Soc. Am.* **55**, 64 (1965).

[5] R. J. Davis, *J. Opt. Soc. Am.* **56**, 837 (1966).

[6] A. Duncanson and R. W. H. Stevenson, *Proc. Phys. Soc.* **72**, 1001 (1958).

[7] W. C. Johnson, *Rev. Sci. Instr.* **35**, 1375 (1964).

[8] D. Fabre and J. Romand, *C. R. Acad. Sci.* **250**, 1226 (1960); *J. de Physique* **22**, 324 (1961).

[9] D. Fabre, Thèse, *Rev. Opt.* **43**, 394 and 504 (1964).

[10] W. R. Hunter, *Optica Acta* **9**, 255 (1962).

[11] L. R. Canfield, G. Hass, and J. E. Waylonis, *Appl. Optics* **5**, 45 (1966).

[12] D. A. Patterson and W. H. Vaughan, *J. Opt. Soc. Am.* **53**, 851 (1963).

[13] R. G. Taylor, T. A. Chubb, and R. W. Kreplin, *J. Opt. Soc. Am.* **55**, 1078 (1965).

[14] P. G. Wilkinson and E. T. Byram, *Appl. Optics* **4**, 581 (1965).

[15] P. Warneck, *J. Opt. Soc. Am.* **55**, 921 (1965).

[15a] D. F. Heath and P. A. Sacher, *Appl. Optics* **5**, 937 (1966).

[16] A. E. Sandström, in *Handbuch der Physik*, ed. S. Flügge (Springer-Verlag, Berlin, 1957), Vol. 30, p. 226.

[17] D. H. Tomboulian, in *Handbuch der Physik*, ed. S. Flügge (Springer-Verlag, Berlin, 1957), Vol. 30, p. 303.

[18] V. E. Cosslett and W. C. Nixon, *X-Ray Microscopy* (Cambridge U. P., 1960).

[19] J. A. Victoreen, *J. Appl. Phys.* **20**, 1141 (1949).

[20] B. L. Henke, R. White, and B. Lundberg, *J. Appl. Phys.* **28**, 98 (1957).

[21] T. A. Chubb, H. Friedman, R. W. Kreplin, R. L. Blake, and A. E. Unzicker, *Mém. Soc. Roy. Sci. Liege*, 5th Ser. IV, 228 (1961).

[22] W. R. Hunter, D. W. Angel, and R. Tousey, *Appl. Optics* **4**, 891 (1965); *see also* Codling et al. *J. Opt. Soc. Am.* **56**, 189 (1966).

[23] H. M. O'Bryan, *J. Opt. Soc. Am.* **22**, 739 (1932).

[24] D. H. Tomboulian and D. E. Bedo, *Rev. Sci. Instr.* **26**, 747 (1955).

[25] D. H. Tomboulian and E. M. Pell, *Phys. Rev.* **83**, 1196 (1951).

[26] N. Astoin and B. Vodar, *J. Phys. Radium* **14**, 424 (1953).

[27] D. Bohm and D. Pines, *Phys. Rev.* **82**, 625 (1951).

[28] D. Pines, in *Solid State Physics*, eds. F. Seitz and D. Turnbull (Academic, New York, 1955); *Revs. Mod. Phys.* **28**, 184 (1956).

[29] J. R. Townsend, *Phys. Rev.* **92**, 556 (1953).

[30] R. W. Woodruff and M. P. Givens, *Phys. Rev.* **97**, 52 (1955).

[31] B. K. Agarwal and M. P. Givens, *Phys. Rev.* **107**, 62 (1957).

[32] D. H. Tomboulian, D. E. Bedo, and W. M. Neupert, *J. Phys. Chem. Solids* **3**, 282 (1957).

[33] W. C. Walker, J. A. R. Samson, and O. P. Rustgi, *J. Opt. Soc. Am.* **48**, 71 (1958).

[34] W. C. Walker, O. P. Rustgi, and G. L. Weissler, *J. Opt. Soc. Am.* **49**, 471 (1959).

[35] N. N. Axelrod and M. P. Givens, *Phys. Rev.* **120**, 1205 (1960).

[36] S. Jeric, J. Robin, and S. Robin-Kandare, *J. Phys. Rad.* **23**, 957 (1962).

[37] S. Robin-Kandare, J. Robin, S. Kandare, and S. Jeric, *Compt. Rend.* **257**, 1605 and 2026 (1963).

[38] W. C. Walker, *J. Phys. Chem. Solids* **24**, 1667 (1963).

[39] W. R. Hunter and R. Tousey, *J. Physique* **25**, 148 (1964).

[40] J. A. R. Samson, *J. Opt. Soc. Am.* **54**, 1491 (1964).

[41] J. A . R. Samson and R. B. Cairns, *Appl. Optics* **4**, 915 (1965).

[42] O. P. Rustgi, *J. Opt. Soc. Am.* **55**, 630 (1965).

[43] D. E. Carter and M. P. Givens, *Phys. Rev.* **101**, 1469 (1956).

[44] H. Kroger and D. H. Tomboulian, *Phys. Rev.* **130**, 152 (1963).

[45] B. Vodar, *Proc. Xth Colloquium Spectroscopium Internationale*, eds. E. R. Lippincott and M. Margoshes (Sparton Books, Washington, D.C., 1963) p. 217.

[46] M. P. Givens, C. J. Koester, and W. L. Goffe, *Phys. Rev.* **100**, 1112 (1955).

[47] Y. Iguchi, *Sci. Light* **13**, 37 (1964).

[48] N. Astoin and B. Vodar, *J. Phys. Radium* **14**, 424 (1953).

[49] A. P. Bradford, G. Hass, M. McFarland, and E. Bitter, *Appl. Optics* **4**, 971 (1965).

[50] P. Metzger and K. Watanabe, private communication, University of Hawaii (1965).

[51] H. Johnson and R. D. Deslattes, *Rev. Sci. Instr.* **36**, 1381 (1965).

[52] L. Harris, *J. Opt. Soc. Am.* **45**, 27 (1955).

[53] U. Hauser and W. Kerler, *Rev. Sci. Instr.* **29**, 380 (1958).

[54] G. Hass, *J. Opt. Soc. Am.* **39**, 532 (1949).

[55] L. Holland, *Vacuum Deposition of Thin Films* (Wiley, New York, 1961).

[56] G. V. Jorgenson and G. K. Wehner, *Trans. 10th National Vacuum Symposium* (Macmillan, New York, 1963).

[57] S. Tolansky, *Multiple Beam Interferometry of Films and Surfaces* (Oxford U. P., London and New York, 1948).

[58] K. Watanabe, *Advances in Geophysics* **5**, 153 (1958).

[59] B. Bates and D. J. Bradley, *Appl. Optics* **5**, 971 (1966).

7

Detectors

The various modes of interaction between radiation and matter provide the underlying principles of all detectors. For vacuum uv radiation, these interactions involve the photoionization of gases, the ejection of photoelectrons from solids, chemical changes, photoconductivity, fluorescence, and thermal effects. Detectors involving most of these principles are discussed below. Thermopiles, bolometers, etc., utilizing the heating effect of radiation, are used primarily to determine the absolute intensity of radiation. Thus a discussion of these instruments will be deferred until Chapter 8.

7.1 PHOTOGRAPHIC PLATES

The gelatin base of all photographic emulsions absorbs radiation in the vacuum ultraviolet, rendering the films insensitive to this spectral region. Two methods are available to sensitize films to vacuum uv radiation. One method is to coat the emulsion with a fluorescent substance that converts the radiation to light of longer wavelengths. The other method, first invented by Schumann [1] in 1892, is to remove practically all of the gelatin.

Fluorescence sensitization of photographic emulsions originated with Duclaux and Jeantet [2], who used thin layers of oil. Oil coated emulsions have also been used by Harrison and Leighton [3] from 3300 to 900 Å. The Eastman Kodak Company has produced a fluorescing lacquer normally used to overcoat their 103 type 0 film. Most of the films sensitized in this manner show comparatively uniform spectral response throughout the range in which the sensitizer absorbs. The sensitizer, of course, must be washed off before development of the film. The Eastman fluorescent lacquer can be removed by washing the film for about one minute in ethylene dichloride. The spectral response of the lacquer has been studied by Johnson, Watanabe, and Tousey [4]. Figure 7.1 shows their results from 1000 to 2800 Å for various thicknesses of the lacquer.

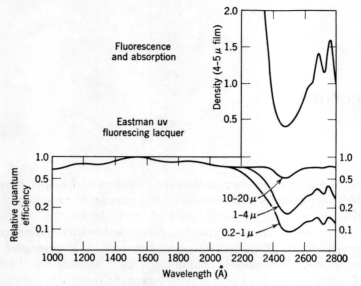

Fig. 7.1 Relative quantum efficiencies for 3 thicknesses of Eastman uv fluorescing lacquer, and the optical density of a film of about 4.5 micron thickness (courtesy F. S. Johnson et al. [4]).

For radiation shorter than 2200 Å, the quantum efficiency is relatively constant. These results have been verified by Lee and Weissler [5], who also obtained the fluorescent emission spectrum of the lacquer. They observed a broad emission band extending from 2900 to 3500 Å with a peak emission at 3150 Å. The fluorescent lacquer was found to be transparent at this wavelength.

Allison and Burns [6] have sensitized films with sodium salicylate while Burrows et al. [7] report that sensitizing with sodium chloride greatly enhanced Ilford *Q* emulsions at wavelengths less than 300 Å. A further property of sodium chloride was that it discriminated against the scattered longer wavelengths.

Coated plates do not appear to have as high a resolving power as the best Schumann plates, but their resolution is still excellent for most applications.

Schumann plates, because of their lack of gelatin, are extremely sensitive to abrasion. This is true of all films and plates where the gelatin content is low. The most important of the commercially produced plates are the Ilford *Q* emulsions, Eastman Kodak SWR (short wavelength radiation) [8], and Kodak Pathé *C*-type emulsions [9]. All of these plates can be handled in a darkroom using a safelight with a No. 6B Wratten yellow filter. The SWR film is more sensitive than the 103-0 uv film below about 1200 Å while the Kodak Pathé SC.7 film is about ten times faster than

170 Å   210 Å

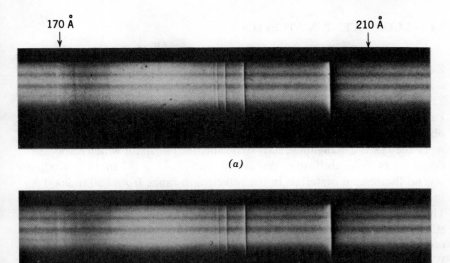

(a)

(b)

Fig. 7.2 The absorption spectrum of helium in the 200 Å region at a helium pressure of 0.25 torr. (*a*) SC7 emulsion. Two minute exposure. Developer D 19 B, development time: 4 minutes. (*b*) SWR emulsion. 20 minute exposure. Developer D 19, 1:1, development time: 2 minutes. (Courtesy K. Codling and R. P. Madden, National Bureau of Standards.)

SWR film. Figure 7.2 shows the helium absorption spectrum in the 200 Å region as recorded by the SWR emulsion and by the SC.7 emulsion.* As can be seen from the Fig. 7.2, the SC.7 emulsion produces a similar photograph with about a factor of 10 less in exposure time.

Because the above plates and films are so sensitive to abrasion, it is advisable to coat the developed plates with a protective lacquer. Kodak Pathé recommend the following:

Nitrocellulose (low viscosity $\frac{1}{3}$ sec)	4 gm
Ethyl acetate	30 gm
Butyl acetate	15 gm
Ethanol	55 gm

This forms a rather low viscosity fluid into which the negatives are completely submerged. After they are dipped in the lacquer the negatives are allowed to dry in a dust free area.

* The author is grateful to Dr. Codling and Dr. Madden for permission to reproduce their excellent absorption spectrum.

7.2 FLUORESCENT MATERIALS

Soon after the first successful photomultiplier was constructed by Zworykin et al. [10] in 1936, the combination of a photomultiplier with a scintillator was commonly used to detect γ rays and nuclear particles. The technique was soon applied to the detection of ultraviolet radiation. Parkinson and Williams in 1949 were probably the first to sensitize a photomultiplier for the detection of vacuum uv radiation [11]. In their work they used a manganese activated willemite phosphor which responded down to 1450 Å. Johnson, Watanabe, and Tousey [4] in 1951 studied several fluorescent materials in the wavelength range from 850 to 2000 Å. Included in this group of materials was the phosphor sodium salicylate. They found that sodium salicylate had one of the highest fluorescent efficiencies (a response which was constant to within ±20 per cent over the range 850–2000 Å), was easily prepared, and was not affected by a vacuum. Watanabe and Inn [12] repeated and confirmed the fluorescent yield experiments on sodium salicylate over the same spectral range. They included one point at 584 Å which lay 15 per cent lower than their mean efficiency value. These ideal qualities shown by sodium salicylate have made it the most commonly used sensitizer for the detection of vacuum uv radiation.

Sodium Salicylate

Probably the earliest investigation of the relative fluorescent yield of sodium salicylate was carried out by Déjardin and Schwégler in 1934 [13]. They reported a constant fluorescent efficiency between 2200 and 3400 Å. Since then, others have confirmed the results for wavelengths longer than 2000 Å [14,15]. For wavelengths shorter than 2000 Å, recent results have shown that the fluorescent efficiency is not always constant as a function of wavelength [16–19]. There appears to be an aging effect which reduces the efficiency more at the shorter wavelengths. This aging effect is possibly due to the vacuum environment, which in most vacuum monochromators contains contaminants such as oil vapors. Allison et al. [20] have studied the aging effect at 1216, 1608, and 2537 Å, taking great care to keep the sodium salicylate samples in a very clean vacuum, and have observed no deterioration at these wavelengths. Although some doubts have arisen as to the constancy of its fluorescent efficiency, sodium salicylate still appears to be the most useful phosphor for detection of vacuum ultraviolet radiation. Consequently its properties will be discussed below in detail.

Preparation. Sodium salicylate is obtained as a very fine crystalline powder which can be dissolved in methyl alcohol. After forming a saturated solution, it is sprayed onto a glass slide or directly onto the glass envelope of a photomultiplier using an atomizer or spray gun. A heat gun is used to blow hot air continually onto the window to facilitate the evaporation of the alcohol. This procedure produces a fine crystalline layer of sodium salicylate. The spraying is continued until the glass surface is all covered and the desired thickness is obtained.

Efficiency versus Thickness.   There appears to be unanimous agreement among the several investigators who have studied the effect of thickness

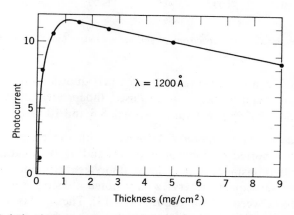

Fig. 7.3   Relation between response and thickness of sodium salicylate layer at 1200 Å (courtesy M. Seya and F. Masuda [21]).

on the fluorescent efficiency that a surface density of approximately 1 mg per cm² will give the optimum response [21–24]. This value appears to be independent of wavelength from 584 to 2200 Å. Figure 7.3 reproduces the work of Seya and Masuda at 1200 Å [21]. It can be seen that the efficiency rises rapidly to a maximum at 1 mg per cm² then falls off very slowly as the thickness increases.

Fluorescent Spectrum and Decay Time.   The fluorescent spectrum of sodium salicylate was first measured by Thurnau [25] who also found that the spectrum was independent of the exciting wavelength between 275 and 2537 Å. Hammann [14] has shown this to be true from 2800 to 3600 Å. Figure 7.4 shows the relative intensity of the emission spectrum as a function of wavelength. The maximum intensity of fluorescence is located at 4200 Å and coincides with the maximum sensitivity of a photomultiplier with an S11 cathode.

The fluorescent decay time of sodium salicylate appears to lie between 7 and 12 nsec. Early measurements by Nygaard and Sigmond [26] in 1961

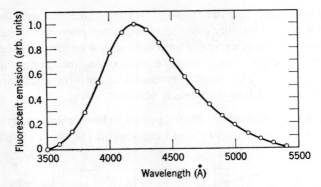

Fig. 7.4 Fluorescent emission spectrum of sodium salicylate (courtesy E. C. Bruner [19]).

gave a value of 12 nsec; however, Nygaard [27] quotes more recent and accurate measurements as having given 7 nsec. Independent measurements by Herb and Sciver [28] give a value between 8.5 and 10 nsec.

Relative Fluorescent Quantum Efficiency. With the exception of the original results reported by Johnson et al. [4] and by Watanabe and Inn [12] regarding the constancy of the relative fluorescent efficiency of sodium salicylate, most subsequent observers have noted a decrease in efficiency for wavelengths between 1600 and 1000 Å [16–19]. The decrease is minimal for a freshly prepared salicylate film which has not been exposed to a typical vacuum monochromator environment for long. Figure 7.5 shows the aging effect for a sample approximately 280 hours old compared to a sample 1 hour old [16]. Where the fresh sample had a constant efficiency

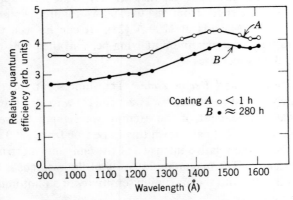

Fig. 7.5 Relative quantum efficiency of sodium salicylate between 900 and 1600 Å. The yield was measured for two different coatings relative to a thermopile. Coating A is less than one hour old while coating B is approximately 280 hours old (Samson [16]).

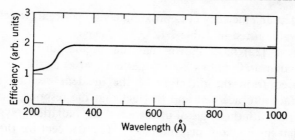

Fig. 7.6 Relative quantum efficiency of sodium salicylate between 200 and 1000 Å.

between 900 and 1216 Å, the 280-hour old sample was 12 per cent lower at 900 than at 1216 Å. Below 1000 Å, the relative efficiency remains constant down to 300 Å, then rapidly decreases by about 50 per cent between 200 and 300 Å. The relative efficiency of sodium salicylate between 200 and 1000 Å is shown in Fig. 7.6 [29]. The efficiency in this case is expressed in terms of the ratio of the photomultiplier output current to the absolute intensity of the incident light (expressed in photons/sec). The absolute intensity being measured with a rare gas ion chamber.

Absolute Fluorescent Quantum Efficiency. As mentioned above, the fluorescent quantum efficiency has been observed to change with age. This fact would seem to make it meaningless to talk about an absolute efficiency. However, even although aging takes place, it probably is possible to determine the intensity of a source to within an order of magnitude using a sodium salicylate coated photomultiplier. Thus it is desirable to determine the absolute efficiency of sodium salicylate as accurately as possible. Table 7.1 lists the various values obtained by different groups. The spread in the results undoubtedly is caused, in part, by samples of different thicknesses and age. However, the spread may also be caused by the different experimental techniques employed. With the exception of Studer [30], all of the investigators have used thin coatings of sodium salicylate on a glass substrate, the sodium salicylate being first dissolved in methyl alcohol. Studer, however, compressed the salicylate to form a plaque about 2 mm thick and then compared the reflected and fluorescent radiation from this plaque with a similar one formed from MgO. He has assumed that all of the fluorescent radiation is emitted from the surface of the salicylate plaque that faces the incident radiation. This assumption is based on the supposition that the sodium salicylate plaque, as that of MgO, has a reflectance approaching 100 per cent. If this assumption is not valid, the absolute efficiency of salicylate will be greater than 60 per cent. However, in a separate experiment, Studer has compared the fluorescence from a sodium salicylate plaque with that from a MgWO$_4$ plaque and has again arrived at an efficiency of approximately 60 per cent, which is about the average value listed in Table 7.1.

Allison et al. [22] have shown that the angular distribution of the fluorescent radiation for films 2-4 mg per cm² varies as the cosine of the angle measured from the direction of the incident exciting radiation. Taking this fact into account and the fact that the response of a photomultiplier varies with the angle of incidence of the radiation, Nygaard [31] has shown that his previous determination of 50 per cent for the absolute efficiency was too small, but the magnitude is unknown. The value of

Table 7.1 Absolute Quantum Efficiency of Sodium Salicylate

Observer	Absolute Efficiency			Layer Thickness	
	2537 Å	1216 Å	304 Å	(mg/cm²)	References
Allison *et al.*	99	94		2–4	[22]
Bruner		62–80	41	5	[19]
Vasseur and Cantin	65	38		2	[18]
Nygaard	50			1–2	[23]
Kristianpoller	64			6	[33]
Inokuchi *et al.*	25			?	[32]
Studer	60			2 mm[a]	[30]

[a] This sample was a pressed plaque 2 mm thick.

25 per cent found by Inokuchi et al. [32] appears to be too low. Thus the main discrepancy at 2537 Å is between the high efficiency found by Allison et al. and the remaining values which lie approximately between 60 and 65 per cent.

From the practical point of view, a value of 65 per cent for the absolute quantum efficiency of sodium salicylate for exciting radiation between 400 and 3400 Å seems to be a suitable compromise. Regardless of any aging effects or of the true quantum efficiency, it is unlikely that the value of 65 per cent will be more than a factor of two in error. It should be noted that Kristianpoller [33] has observed an increase in the fluorescent efficiency of approximately 25 per cent when sodium salicylate is cooled to liquid nitrogen temperature.

Plastic Scintillators

Plastic scintillators form an important group of fluorescent materials in the vacuum uv. Although the fluorescent efficiency of plastic scintillators is much less than that of sodium salicylate and is not even approximately constant with wavelength, such scintillators have the single advantage of having an extremely smooth surface that serves as an ideal substrate for the deposition of thin metallic films. Films that would otherwise oxidize rapidly in air can be over-coated with another transparent metal providing a protective coating. Such combinations mounted onto a photomultiplier can provide excellent narrow band detectors. The combination of a carbon film about 300 Å thick with an aluminum film 1000 Å thick transmits radiation only between 172 and 550 Å [34]. This produces an excellent filter for the He II 304 Å line with a transmittance of approximately 10 per cent. This scintillator-carbon-aluminum combination provides a discrimination between the He 304 and 584 Å lines of approximately 200:1. Conversely, a thin film of tin about 1700 Å thick evaporated onto the plastic scintillator has a transmittance of 20 per cent at the 584 Å line and about a factor of 300 less at the 304 Å line. When metallic films are evaporated onto scintillators, their effective transmittances are increased since they reflect the fluorescent radiation which would otherwise escape detection. Similar results can be achieved using other fluorescent materials [35].

Several types of plastic scintillators are available commercially. The type NE 102* contains scintillation crystals of p-terphenyl (primary solute) and bis-1,4-(2-phenyl-5-oxazolyl)-benzene (secondary solute contracted to read POPOP). The crystals are dissolved in the plastic polyvinyl-toluene. The fluorescent wavelength at maximum emission is 4200 Å, and the fluorescent decay time is 2.2 nanoseconds. NE 103 differs from NE 102 only as regards the nature of the wavelength shifter (secondary solute) and has a slightly longer decay time, namely, 11.3 nanoseconds. The fluorescent spectra apparently are characteristic of the wavelength shifter only. NE 150 contains 10 per cent naphthalene in addition to primary and secondary solutes and exhibits pulse shape discrimination.·

The scintillator Pilot B† also contains p-terphenyl as the primary solute, but uses diphenylstilbene as the secondary solute. The scintillator is quoted to have a decay time of 2.1 nanoseconds with maximum emission at 4070 Å. Figure 7.7 shows the fluorescent spectrum of both Pilot B and NE 102 scintillators. The relative fluorescent efficiency of type NE 102 has been measured as a function of wavelength between 200 and 2000 Å by Samson and Cairns [34]. Their results are shown in Fig. 7.8. The absolute intensity of the exciting radiation was measured between 200 and 1000 Å with use of the rare gas ion chambers. For the region above 1000 Å, the efficiency was measured relative to sodium salicylate. Since the total scintillator solute concentration in the plastic is less than 3 per cent, it is

---

* Nuclear Enterprises Ltd., Winnipeg, Canada.
† Pilot Chemicals, Inc., Watertown, Mass.

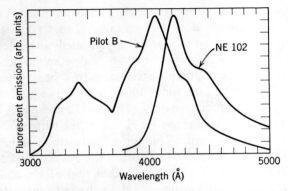

Fig. 7.7 Fluorescent emission spectrum of Pilot B and NE102 plastic scintillators (courtesy Pilot Chemicals, Inc. and Nuclear Enterprises Ltd.).

obvious that most of the radiation will be absorbed by the plastic. Most of the fluorescent excitation will occur on the surface and within the first few hundred angstroms from the surface of the plastic. The efficiency of fluorescence of a plastic scintillator in the vacuum uv should be related to the transmittance of the plastic. The transmittance of polyvinyltoluene has not been measured. However, should it be similar to parlodion or celluloid as shown in Chapter 6 (Fig. 6.7), the minimum in the fluorescent efficiency curve in the vicinity of 800 Å can be explained on the basis that the transmittance of the plastic has a minimum at 800 Å. In fact, the shape of the fluorescent efficiency curve between 300 and 1200 Å is very similar to the shape of the transmittance curve of parlodion over the same wavelength range.

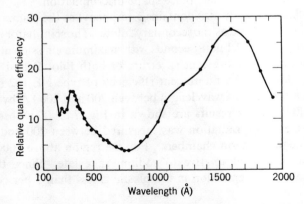

Fig. 7.8 Relative quantum efficiency of NE102 between 200 and 2000 Å (Samson and Cairns [34]).

The fluorescent efficiency of NE 102 was observed to decrease with age, possibly because of sublimation of the scintillator crystals from the surface of the plastic.

Miscellaneous Fluorescent Materials

The relative fluorescent efficiency of diphenylstilbene (DPS) has been measured by Brunet et al. [36] along with tetraphenyl butadine (TPB) and terphenyl (TPh). Their results are shown in Fig. 7.9 relative to sodium

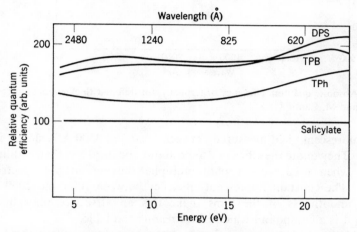

Fig. 7.9 Quantum efficiency of diphenylstilbene (DPS), tetraphenyl butadine (TPB), and terphenyl (TPh) relative to sodium salicylate between 540 and 2500 Å (courtesy M. Brunet et al. [36]).

salicylate. Although the efficiencies of DPS and TPB are much greater than sodium salicylate, they have the disadvantage that they sublime at room temperature in a vacuum and tend to deteriorate when exposed to air for some time. Terphenyl, however, appears to be quite stable. Vasseur and Cantin [18] have measured the absolute fluorescent efficiency of terphenyl and found it to be relatively constant below 1500 Å. Figure 7.10 shows their results from 3000 to 500 Å. The efficiency versus film thickness as measured by Brunet et al. is shown in Fig. 7.11. The terphenyl was deposited using a vacuum evaporator.

A yellow organic phosphor called "liumogen," or 2,2'-dihydroxy-1,1'-naphthaldiazine, has been investigated by Kristianpoller and Dutton [37]. Very smooth vacuum evaporated films can be obtained. They found that the fluorescent efficiency of liumogen increased by about a factor of 2 when cooled to liquid nitrogen temperature. Figure 7.12 shows the fluorescent emission spectrum measured at room temperature and at 80°K. The

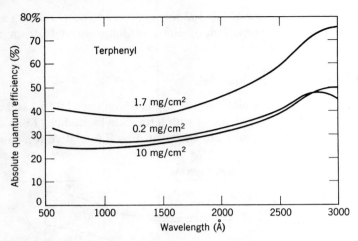

Fig. 7.10 Absolute quantum efficiency of terphenyl for different thicknesses (courtesy J. Vasseur and M. Cantin [18]).

relative fluorescent yield measured between 1000 and 3500 Å is shown in Fig. 7.13. They quote the efficiency to be about one third less than sodium salicylate when used with a photomultiplier having an S-4 spectral response. The constant fluorescent efficiency between 1000 and 2000 Å plus the smoothness of the films make this an attractive phosphor: However, it is a compound which is not readily available.

Inokuchi et al. [32] report that the compound coronene, $C_{24}H_{12}$, exhibits a constant fluorescent efficiency between 1500 and 3000 Å and has a higher efficiency than sodium salicylate. Unfortunately, the maximum

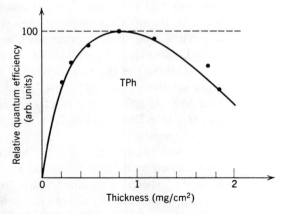

Fig. 7.11 Relative quantum efficiency of terphenyl as a function of film thickness (courtesy M. Brunet et al. [36]).

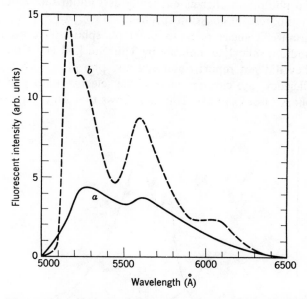

Fig. 7.12 Fluorescent emission spectrum of liumogen (*a*) at room temperature (*b*) at liquid nitrogen temperature (courtesy N. Kristianpoller and D. Dutton [37]).

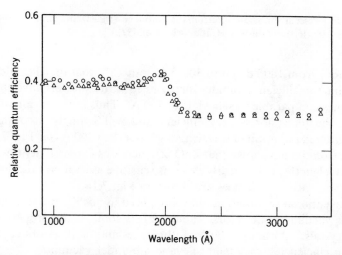

Fig. 7.13 Quantum efficiency of liumogen relative to sodium salicylate between 1000 and 3500 Å (courtesy N. Kristianpoller and D. Dutton [37]).

of its fluorescent emission, as that of liumogen, does not match the spectral response of a photomultiplier as efficiently as sodium salicylate. Figure 7.14 gives the fluorescent emission spectrum.

The quantum efficiencies of many other phosphors have been studied relative to sodium salicylate, notably by Thurnau [25] and Conklin [38]. Although they did not report observing any phosphor with a constant quantum efficiency, use can be made of the selective nature of some of these phosphors. For example, Thurnau shows that $Zn_3(PO_4)_2$:Mn has a

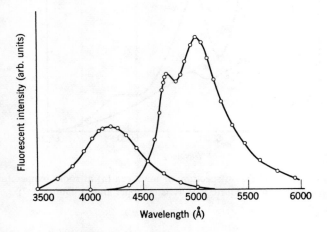

Fig. 7.14 Fluorescent emission spectrum of coronene (right) compared to that of sodium salicylate (left). (Courtesy H. Inokuchi et al. [32].)

low efficiency from 1800 down to 700 Å (about 25% that of salicylate), at which point the efficiency rapidly increases towards shorter wavelengths and becomes comparable to salicylate at 200 Å. Thus in the region of 200 to 400 Å, the phosphor is very efficient and will strongly discriminate against scattered radiation of wavelength greater than 600 Å (see Fig. 7.15). Similarly, Conklin has shown that ZnO:Zn increases in sensitivity towards shorter wavelengths, being nearly twice as sensitive as sodium salicylate at 400 Å. The relative efficiencies are shown in Fig. 7.16.

The conventional scintillation phosphors used in nuclear physics can all be used as vacuum uv detectors. Table 7.2 lists some of the properties of several crystals. Although both anthracene and stilbene have high fluorescent efficiencies, they tend to evaporate under vacuum.

Most materials fluoresce to a greater or less extent when irradiated by vacuum uv radiation. This fact can often cause trouble when measuring the transmittance of lithium fluoride and calcium fluoride windows. Since

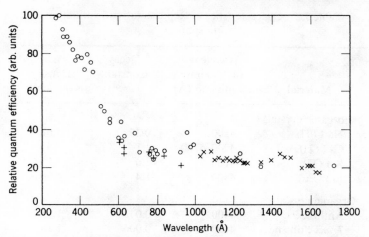

Fig. 7.15 Quantum efficiency of $Zn_3(PO_4)_2$: Mn relative to sodium salicylate between 200 and 1800 Å (courtesy D. H. Thurnau [25]).

these windows are so commonly used, care must be taken in any experiment to determine the possible error introduced by their fluorescence. Sometimes it is of interest to detect very weak signals of visible light in a background of vacuum uv radiation. However, the glass envelope of a photomultiplier tube fluoresces slightly and vacuum uv radiation will be detected. Thus it is desirable to coat the photomultiplier window with a material which does not fluoresce. A thin film of collodion is suitable in

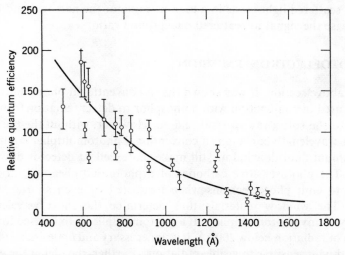

Fig. 7.16 Quantum efficiency of ZnO:Zn relative to sodium salicylate between 400 and 1800 Å (courtesy R. Conklin [38]).

Table 7.2   Characteristics of Representative Scintillation
Phosphors[a]

Material	Wavelength of Maximum Emission (Å)	Decay Constant (μsec)	Density (gm/cc^3)
Inorganic crystals			
Na I (Tl)	4100	0.25	3.67
Cs I (Tl)	4200–5700	1.1	4.51
K I (Tl)	4100	1.0	3.13
Li I (Eu)	4400	1.4	4.06
Organic crystals			
Anthracene	4400	0.032	1.25
Trans-Stilbene	4100	0.006	1.16
Plastic phosphors	3500–4500	0.003–0.005	1.06
Liquid phosphors	3500–4500	0.002–0.008	0.86

[a] Courtesy Harshaw Chemicals, Inc.

this case. Since most materials fluoresce in the blue, the use of red or yellow filters can accomplish the same effect. Conversely, if the vacuum uv detection system is observing a background of radiation in the red end of the spectrum, such as a furnace, the use of a narrow band filter at the maximum wavelength of the fluorescent emission can be used. If considerable scattered light exists in a monochromator the narrow band filter will increase the signal to scattered background ratio.

7.3 PHOTOELECTRON EMISSION

In the above section, it was shown that a conventional photomultiplier could be used in conjunction with a phosphor to detect radiation from the visible into the soft x-ray spectral region. However, without the use of a phosphor wavelength converter, a conventional photomultiplier equipped with a lithium fluoride window still makes an excellent detector down to 1050 Å. The photosensitive cathodes of a photomultiplier will, in fact, continue to emit photoelectrons when irradiated by even shorter wave lengths. The window materials thus determine the short wavelength sensitivity of any photomultiplier. If a photomultiplier is to be used for the detection of radiation below 2000 Å, it is unnecessary and undesirable to use photo-cathodes sensitive to visible radiation. Further, the higher the work function of the photocathode material, the smaller will be the thermal dark current. A family of new photocathodes have been introduced recently that are relatively insensitive to radiation of wavelengths longer than about 3000 Å [39]. Since this is approximately the wavelength limit of the solar radiation which reaches the earth's surface, the term "solar blind" has been applied to photomultipliers that use these cathodes. With a lithium fluoride window, these multipliers are excellent detectors from 1050 to 2500 Å. Typical cathodes are made from cesium telluride, rubidium telluride, cesium iodide, and copper iodide. Figure 7.17 illustrates the

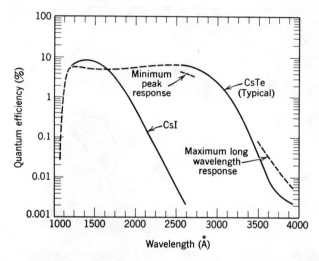

Fig. 7.17 Typical quantum efficiencies of EMR photomultipliers with photocathodes of cesium iodide and cesium telluride (courtesy Electro-Mechanical Research, Inc.).

quantum efficiency of two EMR photomultipliers* equipped with LiF windows and cathodes of cesium telluride (Model 541F, 18 stage) and cesium iodide (Model 541G, 18 stage). The manufacturer quotes dark currents at a gain of 10^6 to be 3×10^{-12} A for cesium iodide cathodes and 2×10^{-11} A for cesium telluride cathodes. Such low dark currents enable this type of multiplier to detect weaker signals in the range 1050 to 2400Å than do the phosphor coated multipliers. Thin metallic films of gold have also been used for photo-cathodes, however, the photoelectric yield of a pure metal is somewhat lower than the alkali iodides and tellurides in this wavelength region.

To utilize solar blind detectors below 1050 Å no window material can be used. This is a disadvantage for cathodes of the alkali iodide and telluride variety since, on exposure to air, there is a general decrease in their photo

* Electro-Mechanical Research Inc., Princeton, New Jersey.

electric yields. However, when maintained in a good vacuum, their yields are considerably higher than any metal. The cathodes are prepared by evaporating extremely thin films ($\sim$1000 Å) onto a smooth metal substrate If the films are too thick or if they are exposed to a very intense source of radiation, their response will not be linear with respect to the incident intensity. Thus great care must be exercised when using nonconductors for photocathodes. Figure 7.18 shows the photoelectric yield of CsI and RbI

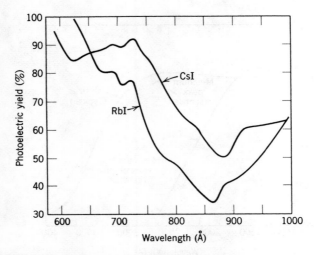

Fig. 7.18 Photoelectric yield of CsI and RbI as a function of wavelength (courtesy P. H. Metzger and K. Watanabe [40]).

between 1000 and 600 Å as obtained by Metzger [40]. The data were taken with the radiation incident at 45° to the cathodes, resulting in a yield some 20 per cent higher than that obtained at normal incidence. Metzger and Watanabe have found that all of the alkali halides have a yield in excess of 20 per cent between 600 and 1000 Å, while most have yields in excess of 40 per cent. The photoelectric yield as defined here is the number of *photoelectrons ejected per incident photon* (η). This is a practical definition rather than the more fundamental definition of *photoelectrons per absorbed photon* (γ) that takes into account the reflectance of the cathode. The practical definition will be used here unless otherwise stated. The photoelectric yield of strontium and lithium fluoride are shown in Figs. 7.19 and 7.20, respectively. The structure in the yield curves is indicative of the energy band structure of the insulators.

The use of a metal cathode is perhaps more convenient in the "windowless" region of the vacuum uv. A normal metal surface which has been

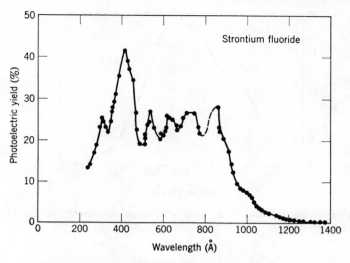

Fig. 7.19 Photoelectric yield of SrF$_2$ as a function of wavelength.

exposed to air has a very stable photoelectric yield. Although not so sensitive as the alkali halides, most metals have yields in excess of 10 percent between 1000 and 400 Å. A clean metal surface formed under ultra high vacuum conditions has a much lower yield. The oxide layer which forms when a metal is exposed to air is partially responsible for the increase in yield. A highly polished surface has a greater yield than a matt

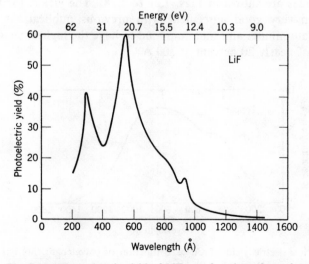

Fig. 7.20 Photoelectric yield of LiF as a function of wavelength.

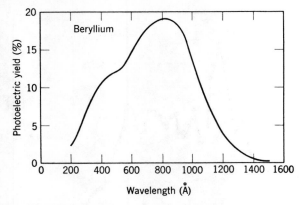

Fig. 7.21 Photoelectric yield of Be as a function of wavelength, normal incidence (Cairns and Samson [41]).

surface. This is probably due to the fact that most photoelectrons produced within crevices are recaptured by the metal before they can escape. This fact can be made use of when it is desirable to suppress photoelectric emission from a surface. For example, the maximum efficiency of a gold black surface is only about 4 percent [16].

The photoelectric yields of most metals have been measured by Cairns and Samson [41] from 200 to 1500 Å. The samples were all highly polished and cleaned with methyl alcohol before measurement. This simple procedure gave very reproducible results. The yields of some commonly used photocathodes are shown in Figs. 7.21 to 7.28. The yields, for the most part, are in very good agreement with previous publications [42–44]. Beryllium, aluminum, and tantalum tend to have the highest yields with a maximum of nearly 20 percent at 800 Å.

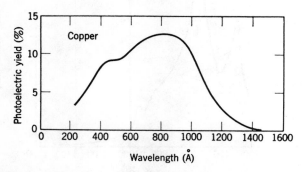

Fig. 7.22 Photoelectric yield of Cu as a function of wavelength, normal incidence (Cairns and Samson [41]).

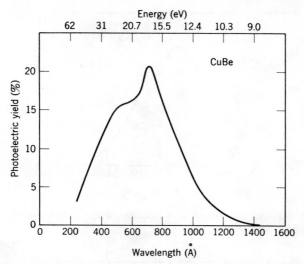

Fig. 7.23 Photoelectric yield of CuBe as a function of wavelength, normal incidence (Cairns and Samson [41]).

All of the above cathodes can be used as simple diodes as illustrated in Fig. 7.29. The photoelectric current saturates with a collector voltage typically between 10 to 100 volts depending on the geometry of the collector electrode. It is usually better to measure the loss of photoelectrons from the cathode rather than trying to collect all of the ejected photoelectrons. As an example of the magnitude of the photoelectric current, consider a flux of 10^6 photons per sec incident on a cathode whose photoelectric yield is 10 percent: 10^5 electrons per sec will be ejected;

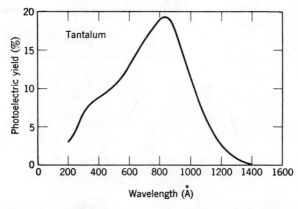

Fig. 7.24 Photoelectric yield of Ta as a function of wavelength, normal incidence (Cairns and Samson [41]).

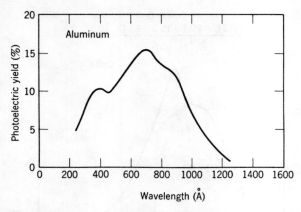

Fig. 7.25 Photoelectric yield of Al as a function of wavelength, normal incidence (Cairns and Samson [41]).

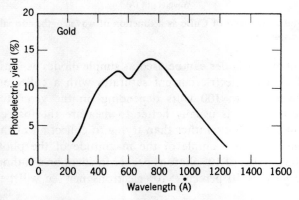

Fig. 7.26 Photoelectric yield of Au as a function of wavelength, normal incidence (Cairns and Samson [41]).

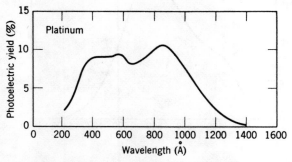

Fig. 7.27 Photoelectric yields of Pt as a function of wavelength, normal incidence (Cairns and Samson [41]).

230

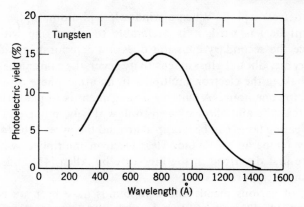

Fig. 7.28 Photoelectric yields of W as a function of wavelength, normal incidence (Cairns and Samson [41]).

that is, the current will be 1.6×10^{-14} A. A current of this size can be measured with standard microammeters; however, it is approaching the noise level. For increased sensitivity, the glass envelope of a standard photomultiplier can be removed, allowing the radiation to strike the first dynode directly. Although the noise level increases considerably, there is a net overall gain in signal of perhaps one thousand, allowing a minimum detectable signal of about 10^3 photons per sec to be detected for an initial yield of 10 percent. To detect signals less than 10^3 photons per sec, pulse-counting techniques should be used. Photon fluxes of a few photons per sec

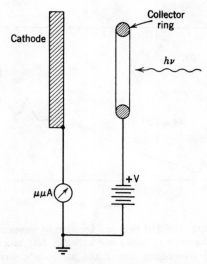

Fig. 7.29 uv detector using a simple metal cathode.

can then be detected [45]. When a windowless photomultiplier (i.e., an electron multiplier) is used, it is preferable to select one with Cu-Be dynodes since the secondary emission does not deteriorate rapidly when exposed to dry air. Should a loss in sensitivity occur, the gain can be partially restored by baking the electron multiplier in an atmosphere of oxygen at about 300°C for one hour. A suitable arrangement is to place the multiplier in a pyrex tube and allow oxygen to flow slowly in one end and out the other. Heating tapes can be wrapped around the pyrex tube to provide the necessary temperature. Windowless electron multipliers with Cu-Be dynodes were first described in the literature by Allen [45] and are now available commercially.

The yields of various metals and compounds have been measured by Lukirskii et al. [46,47] and by Rumsh et al. [48] between 1 and 113 Å, However, these data were of a preliminary nature. Later investigations by Lukirskii et al. [49] suggest that these earlier values were probably too high. To obtain the necessary sensitivity to detect the weak photoelectric signals in this wavelength range, they used their samples as the first dynode of an electron multiplier and employed pulse counting techniques to determine the total number of electrons ejected from the first dynode. The absolute intensity of the incident radiation was determined with a specially constructed Geiger counter. Their more recent data [49] are shown in Figs. 7.30 to 7.32 for BeO, LiF, MgF₂, SrF₂, and CsI. With the exception of BeO, the photo-cathodes were prepared by vacuum evaporation of the compounds onto glass substrates precoated with a thin layer of aluminum. The thickness of the fluoride films was about 2500 Å, while the CsI film

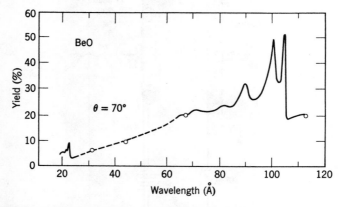

Fig. 7.30 Photoelectric yield of BeO as a function of wavelength for radiation incident at an angle $\theta°$ to the normal. Solid line indicates data taken with a continuum light source; open circles represent data taken at discrete wavelengths (courtesy A. P. Lukirskii et al. [49]).

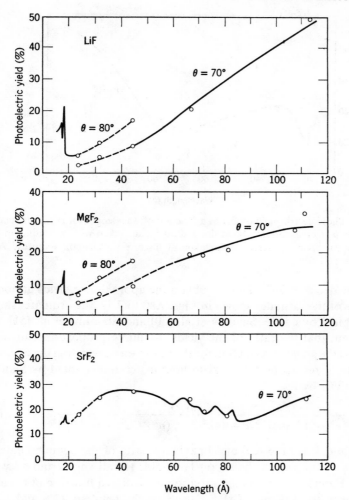

Fig. 7.31 Photoelectric yield of LiF, MgF$_2$, and SrF$_2$ as a function of wavelength for radiation incident at an angle $\theta°$ to the normal. Solid line indicates data taken with a continuum light source; open circles represent data taken at discrete wavelengths (courtesy A. P. Lukirskii et al. [49]).

was approximately 5500 Å thick. The BeO cathode was formed by evaporating Be onto a polished substrate of tungsten or molybdenum and then oxidizing the film by creating a glow discharge in oxygen in the vicinity of the cathode. The yields were measured with the radiation incident at 70° to the normal. Rumsh et al. [48] have shown that in the soft x-ray region, the ratio of the yields measured at an angle of $\theta°$ to the normal to that measured at normal incidence is approximately equal to

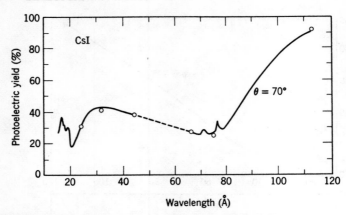

Fig. 7.32 Photoelectric yield of CsI as a function of wavelength for radiation incident at an angle $\theta°$ to the normal. Solid line indicates data taken with a continuum light source; open circles represent data taken at discrete wavelengths (courtesy A. P. Lukirskii et al. [49]).

sec θ provided $\theta < 80°$. Thus to obtain the approximate yields at normal incidence of the cathodes shown in Figs. 7.30 to 7.32, the ordinates should be multiplied by cos θ. Savinov et al. [50] and Stanford et al. [51] have pointed out that the ratio of the yields should be proportional to sec θ_r, where θ_r is the angle of refraction for the radiation penetrating the cathodes. The angle of refraction for an absorbing material as quoted by Stanford et al. is given by

$$\sin \theta_r = \frac{(2 \sin \theta)^{\frac{1}{2}}}{\{n^2 - k^2 + \sin^2 \theta + [4n^2k^2 + (n^2 - k^2 - \sin^2 \theta)^2]^{\frac{1}{2}}\}^{\frac{1}{2}}}, \quad (7.1)$$

where n and k are the real and imaginary parts of the complex index of refraction. Using (7.1), the secant law holds for all angles and all wavelengths. Figure 7.33 shows the variation in yield as a function of the angle of incidence for LiF and CsI for wavelengths between 23.6 and 113 Å [49]. These curves are characteristic of most photocathodes. The maxima of the curves occur at angles of incidence where the increase in reflectance becomes greater than the increase in the yield. For longer wavelengths, the maxima tend to occur at smaller angles of incidence. At 584 Å, the maximum yield of aluminum occurs at about 60°, whereas at 1216 Å, there is no maximum [44]. Table 7.3 tabulates the photoelectric yield of several cathodes at a few selected wavelengths between 23 and 113 Å, measured at an angle of 70° to the normal. Lukirskii et al. [49] estimate the accuracy of their yield measurements to be about ±15 percent for cathodes other than BeO. They found the yield of BeO varied depending on the amount of oxidation of the beryllium film.

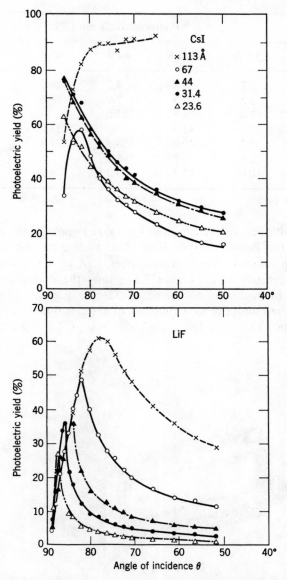

Fig. 7.33 Photoelectric yield of CsI and LiF as a function of the angle of incidence for selected wavelengths: ×, 113 Å; ○, 67 Å, ▲, 44 Å; ●, 31.4 Å; △, 23.6 Å (courtesy A. P. Lukirskii et al. [49]).

Table 7.3 Photoelectric Yields of Various Compounds Measured
at an Angle of 70° to the Normal[a]

Photocathode	Photoelectric Yield (%)				
	23.6 Å	31.4 Å	44 Å	67 Å	113 Å
LiF	2.7	4.8	8.2	20.0	48
MgF_2	3.3	6.3	9.0	19.5	33
SrF_2	17.0	24.5	27.0		23
KCl	29.5	42.0		13.3	22
CsI	31.5	40.0	38.5	27.5	91
BeO	2.8	5.7	8.8		19.3
Au	10.0	12.5	9.0	4.2	5.4

[a] Lukirskii et al. [49].

Electron multipliers for general and specialized applications have been developed by the Bendix Corporation. One such type, the continuous resistance strip magnetic electron multiplier, can be made with an extremely large cathode area. Figure 7.34 shows the multiplier used by Hinteregger in his rocket monochromator [52–54]. Figure 7.35 shows the sensitivity of the multiplier as a function of the position of illumination on the cathode. This type has a cathode area of 2.5 × 9 cm. At the

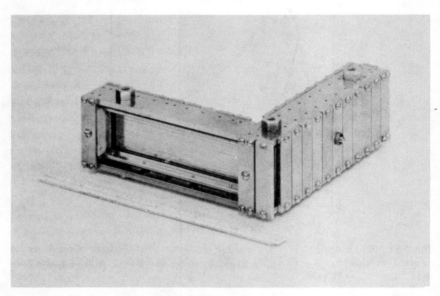

Fig. 7.34 Bendix magnetic electron multiplier (courtesy H. E. Hinteregger [52]).

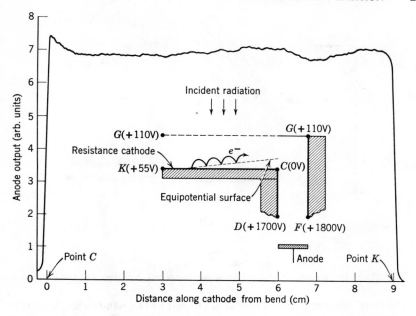

Fig. 7.35 Anode output of Bendix multiplier shown in Fig. 7.34 as a function of the position of illumination on a resistance cathode. The effect of maintaining a potential drop of $+55$ volts along this cathode is shown schematically in the inset. The resistant strips have the following typical values: $KC = 24$ megohm, $CD = 6.8$ megohm, $GF = 9.0$ megohm (courtesy H. E. Hinteregger, unpublished results).

other extreme is the channel multiplier which consists of a thin tube coated internally with a high resistance material $\sim 10^8$ ohms [55–57]. Electrodes are attached at each end, and a field of approximately 2000 V is applied. The channel multiplier can be made as small as 0.2 mm in diameter by about 10 mm in length. The optimum length to bore ratio is about 50:1. Gains of the order of 10^6 can be obtained.

Another technique for multiplying the photoelectric signal from a metal cathode has been used by Lincke and Wilkerson [58]. This method requires the acceleration of the primary photoelectrons to energies of approximately 15 kV. The high energy electrons then impinge upon an aluminum film-scintillator combination, with the fluorescent radiation detected by a conventional photomultiplier (see Fig. 7.36). This is essentially the same principle as described by Daly [59] and Schönheit [60] for the detection of either positive or negative ions. The main advantage of this method is that the dynodes of the multiplier are not subjected to any deterioration of secondary emission due to exposure to air. However, the fact that such high voltages are present is a drawback in many cases;

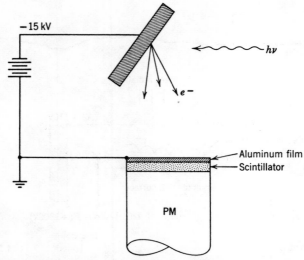

Fig. 7.36 Detector based on the scintillation produced by high energy electrons.

also the dark current is much higher than that of the windowless electron multiplier.

A somewhat different approach has been taken by Samson and Cairns [61] whereby the primary photoelectron emission current is increased by directing the incident radiation at grazing angles to the cathode. The principle of the method is as follows. As already mentioned, the photo-electric yield of a metal increases with the angle of incidence θ of the radiation [48,61,62]. However, if the yield is defined in the practical sense as electrons per incident photon (η), the yield only increases to the point where the reflectance is sufficiently high that most of the incident photons are reflected, and the yield then drops. When the more funda-mental definition is used, namely, electrons per absorbed photon (γ), it is found that the yield continually increases. That is, of the photons absorbed, more produce photoelectrons when the angle of incidence increases. To illustrate the magnitude of the reflectance and the increase in γ, Fig. 7.37 shows the data for tungsten at a wavelength of 584 Å. At an angle of incidence of 80°, the reflectance is nearly 70 per cent, and γ has increased by a factor of 2.4. Thus, it is clear that if the reflected 70 per cent could be reused to produce more photoelectrons to the point where the incident light was completely absorbed, then the total emission current should be increased by approximately 2.4. Several different geometrical forms can be used to recapture the reflected radiation [61]. However, the most useful and compact arrangement is that of a polygon. Figure 7.38 shows a schematic of the cathode while Fig. 7.39 shows a photograph of an actual

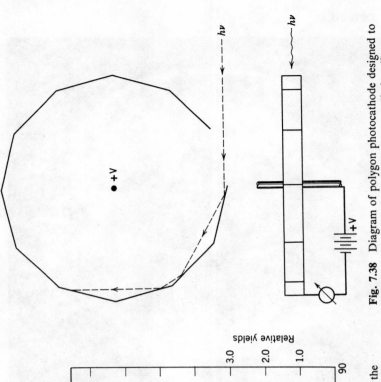

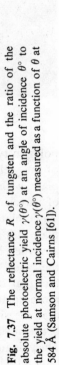

Fig. 7.38 Diagram of polygon photocathode designed to give total absorption of the incident radiation (Samson and Cairns [61]).

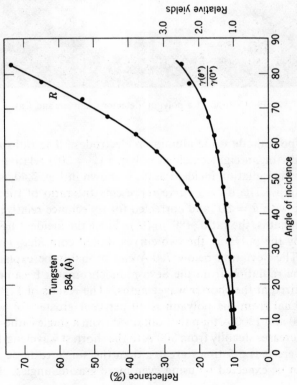

Fig. 7.37 The reflectance R of tungsten and the ratio of the absolute photoelectric yield $\gamma(\theta°)$ at an angle of incidence $\theta°$ to the yield at normal incidence $\gamma(0°)$ measured as a function of θ at 584 Å (Samson and Cairns [61]).

Fig. 7.39 Photograph of a polygon detector (Samson and Cairns [61]).

polygon type cathode using aluminum electrodes. The ratio $\eta(80°)/\eta(0°)$ of the yields for an eighteen sided polygon ($\theta = 80°$) relative to a single plate, with the radiation incident at 0°, is shown in Fig. 7.40 as a function of wavelength. The dotted curve represents the ratio of the yields of a single plate with $\theta = 80°$ and corrected for reflectance relative to a single plate at 0°, that is, the ratio $\gamma(80°)/\eta(0°)$. Since the incident light is totally absorbed by the polygon, the two curves should coincide. This is nearly the case. The deviation below 700 Å can probably be explained by the fact that the radiation from the Seya monochromator became more and more polarized at the shorter wavelengths. The results of Fig. 7.40 show that the signal from the polygon is 40 per cent greater for wavelengths between 800 and 1300 Å than that obtained from a single cathode ($\theta = 0°$). The gain increases steadily from 800 Å to the shortest wavelength measured (209 Å) where it is 5.5 times greater than the single cathode. A further increase can be expected by using even more grazing angles. Figure 7.41

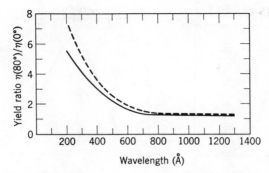

Fig. 7.40 Ratio of the practical yields for an 18-sided polygon ($\theta = 80°$) to that of a single plate ($\theta = 0°$). The dotted curve represents the ratio of the absolute yield of a single plate for $\theta = 80°$ to the practical yield of a single plate for $\theta = 0°$.

shows the photoelectric yield η of an aluminum polygon ($\theta = 80°$) and the yield of a single plate used in the polygon ($\theta = 0°$). The polygon has a yield in excess of 20 per cent between 200 and 1000 Å.

The polygon-type cathode is especially useful for short wavelengths and can be used as the cathode of an electron multiplier or as a simple diode, or it can be used to produce electrons for scintillation detectors.

The photoelectric effect has been used to activate Geiger counters for the visible and ultraviolet regions [63–65]. In the vacuum uv, Turner [66,67] has used copper iodide as the photoemissive surface and fluorite windows. The counter is operated in the proportional region to insure a linear response over a wide range of intensities. Methane was used as the gas

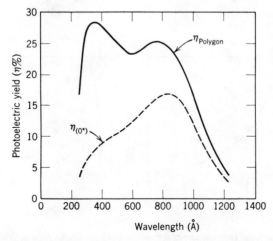

Fig. 7.41 Photoelectric yield of an aluminum polygon ($\theta = 80°$) compared to the yield of a single aluminum plate used in the polygon ($\theta = 0°$) (Samson and Cairns [61]).

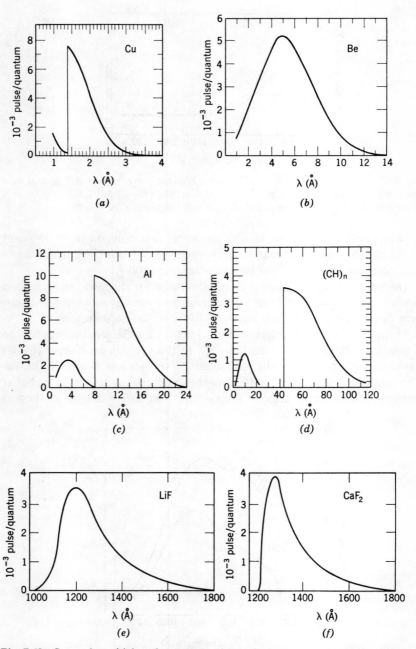

Fig. 7.42 Spectral sensitivity of apparatus with BeO photocathode and with filters of (*a*) Cu, 16 mg. cm^{-2} (*b*) Be, 10 mg. cm^{-2} (*c*) Al, 1.4 mg. cm^{-2} (*d*) (CH)$_n$, 0.5 mg. cm^{-2} (*e*) LiF, 0.5 mm (*f*) CaF$_2$, 0.5 mm (courtesy Efremov et al. [71]).

filling at a pressure of 12 mm Hg. No absolute efficiency for the counter was given, but from the knowledge of the photoelectric yield of CuI [68,69] and the transparency of LiF windows, an efficiency of 5 per cent seems reasonable. Solar blind multipliers using CuI cathodes are also capable of detecting single electron pulses when used in a counting mode [70]. The multipliers have the advantage of a large range in sensitivity and, of course, can be used in conjunction with dc amplifiers. The CuI photon counter is more economical to use since it does not require a linear amplifier and discriminator for operation.

Photoelectric detectors can be combined with the various filters described in the last chapter to produce narrow band detectors throughout the vacuum uv region. The spectral sensitivity of an electron multiplier with a BeO photo-cathode combined with various filters has been investigated by Efremov et al. [71]. Their results are shown in Figs. 7.42 (a)–(f).

Image converter tubes are commercially available from Electro-Mechanical Research, Inc. that are sensitive to radiation down to 1050 Å. The converter consists simply of a curved LiF window coated internally with a semitransparent film of cesium telluride. The vacuum uv image is focused onto the photocathode and is converted into a focused cloud of electrons. The electrons are accelerated through approximately 16 kV and impinge on a phosphor screen which fluoresces with a maximum output at 5600 Å. With simple electrostatic focusing, the converter has a resolution of 25 line pairs per mm on its axis, but considerably less off its axis. There is an overall photon gain of approximately 50 at 2537 Å and probably at least 25 at 1216 Å. Thus the image converter can be used as an amplifier to detect weak uv signals.

Westinghouse has successfully constructed a television camera tube known as the Uvicon that uses a LiF lens focusing system [72]. This tube can resolve approximately 10 line pairs per mm. The Uvicon tube was developed for space research and at present has not been used in spectroscopy. It should find application in the "instantaneous" monitoring of emission lines from fluctuating or transient sources. The image orthicon television camera tube has been used in the visible region of the spectrum to study transient emission sources [73]. It would appear that the orthicon tube could be used to detect radiation of wavelengths less than 1000 Å. It has never been used in this wavelength region. However, the following brief description of the operation of the orthicon tube shows in principle that the tube should be capable of detecting vacuum uv radiation.

A schematic diagram of the image orthicon tube is shown in Fig. 7.43. The incident radiation is focused onto a photosensitive surface, and the photoelectrons released are accelerated by an axial electric field and are focused onto a thin specially prepared glass target by an axial magnetic

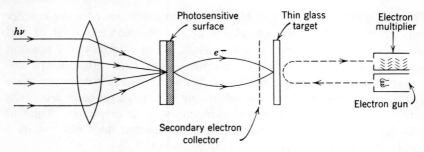

Fig. 7.43 Schematic diagram of the image orthicon.

field. Secondary electrons are ejected from the target and are collected by the fine wire mesh screen. The target is now left with a positively charged image on the side facing the photosensitive cathode. An electron beam on the other side of the target is continually scanning the back surface of the target. A negative charge is deposited on this side equal to the positive charge on the opposite side, while electrons in the beam, in excess of the number needed to neutralize the positive charge, are reflected and return to be detected and amplified by the electron multiplier. As the electron beam scans the target, the anode current of the multiplier is modulated in phase with the positive electrostatic image on the target.

To utilize the orthicon tube for vacuum uv detection, the lens and the glass window with the photocathode must be removed. An auxiliary focusing system is necessary such as a vacuum spectrograph. If the glass target is situated tangentially to the Rowland circle, the spectral lines will be approximately focused onto the target. It may be possible to use a slightly curved target to fit the curvature of the Rowland circle. The photoelectric emission from thin insulating films is quite high (see Figs. 7.19 and 7.20), and it is to be expected that this will be the case for the presently used targets. However, a thin layer of LiF or MgF_2 could, if necessary be deposited onto the target. At any rate, the direct imaging of the spectral lines onto the target will produce a positive electrostatic image which should then be detectable in the conventional manner. It is possible that the amount of the positive charge will not be sufficient to be detected. However, by increasing the time between scans by the electron beam, it should be possible to build up a detectable charge before migration of the charges destroys the image.

7.4   PHOTOIONIZATION

The photoionization efficiency of many gases approaches 100 per cent, especially at wavelengths much shorter than the ionization threshold.

The terms "efficiency" or "yield" (γ) are used interchangeably and are defined as the number of ion pairs produced per photon absorbed. When a photon is energetically capable of doubly ionizing an atom, the yield may be greater than 100 per cent. This occurs, for example, at wavelengths shorter than 371.88 Å in the case of xenon. Detectors utilizing the principle of photoionization have the advantage of high sensitivity, low noise level, and complete insensitivity to wavelengths longer than the ionization potential of the gas. Operating in a dc mode, the detectors are known as ionization chambers and gas-gain ionization chambers. When a pulse mode is used, the detectors are known as proportional counters and Geiger counters (or photon counters).

Ionization Chambers

The number of ions produced in an ionization chamber depends on the number of photons absorbed by the gas and on the photoionization yield of the gas. The number of photons absorbed by a gas is given by the Lambert-Beer law as

$$\text{number of photons absorbed} = I(1 - e^{-\sigma n L}), \tag{7.2}$$

where I is the intensity of the radiation (measured in photons per sec), entering the ion chamber of length L, n is the number of atoms or molecules per cm^3, and σ is the total absorption cross section of the gas. If the transmittance of the window material is T, and I_0 is the intensity of the incident radiation, then $I = TI_0$. Thus the number of primary ions produced per second is given by

$$\frac{i}{e} = TI_0\gamma(1 - e^{-\sigma n L}), \tag{7.3}$$

where i is the electric current in amperes, and e is the electronic charge. When the incident light is completely absorbed by the gas, (7.3) becomes

$$\frac{i}{e} = TI_0\gamma. \tag{7.4}$$

When all the ions formed are collected (7.4) represents the total current flowing in the ion chamber, provided the applied voltage is not too large. A typical ion current vs. voltage characteristic curve is shown in Fig. 7.44 for nitric oxide at a pressure of 20 torr. The ion current rapidly saturates at a few volts then stays constant for nearly 150 volts before ion multiplication sets in. Gains of up to 100 can be achieved around 700 volts. The actual values quoted are, of course, dependent on the geometry of the ion chamber and on the gas pressure. Ion multiplication or gas gain starts when an electron gains sufficient energy to cause secondary ionization. In

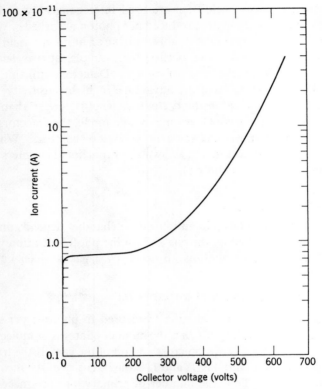

Fig. 7.44 Plateau and gas gain characteristics of a nitric oxide ion chamber such as that shown in Fig. 7.45.

the case of NO, this requires 9.15 eV. This energy must be acquired by the electron in traversing a collision-free path in an electric field E. The electron experiences elastic collisions until it gains sufficient energy to excite the atoms or molecules. As the applied voltage increases, the electron will gain sufficient energy to ionize the gas. This is the start of the Townsend avalanche.

The most effective design for an ion chamber when used in the gas-gain mode is a cylindrical electrode with a thin central wire. The electric field of such a coaxial arrangement is given by

$$E = \frac{V}{r \ln (a/b)}, \tag{7.5}$$

where V is the applied voltage between the cylinder of radius a and the wire of radius b and E is the electric field at any point a distance r from the center of the wire. Thus, the field increases rapidly as $r \rightarrow b$. The collector voltage should be applied such that the electrons and not the ions are driven over to the central wire from which the signal is obtained. The reason for this is that electrons have a longer mean-free path than ions and are responsible for the start of the Townsend discharge. The classical mean-free path of an electron is about four times greater than that of an atom. If the polarity of the applied field is reversed, considerably higher voltages are necessary to give a similar ion multiplication.

When the ion chamber is used in the plateau region, it is normally immaterial whether the ions are collected or the electrons. However, it

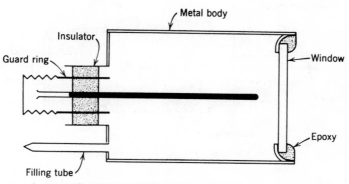

Fig. 7.45   Cylindrical ion chamber.

is possible that photoelectrons can be ejected from the walls of the ion chamber either by radiation entering into the ion chamber or by fluorescence from the electron excited gas. In either case, the photoelectrons would constitute a small error in the absolute determination of the intensity of the incident radiation. Thus it is preferable to collect ions.

In the construction of an ion chamber it is important that there should be no leakage current between the ion collector electrode and the applied voltage electrode. The two electrodes should be separated from each other with a grounded guard ring. Figure 7.45 shows the construction of a cylindrical ion chamber which can also be used in the gas gain mode. The construction of ion chambers is described in detail in Chapter 8.

Photon Counters

The counting action of a photon counter is identical to that of the Geiger counter ordinarily used to detect γ-rays and x-rays. The combination of the ionization threshold and window transmittance cut-off determines the spectral range of sensitivity. For most photon counters when the incident radiation creates a photoelectron, the counter registers a pulse. Thus the efficiency of such a counter is simply the product of the ionization efficiency and the transmittance of the window, assuming the gas completely absorbs all of the radiation. However, if the counter is filled with an electro-negative gas such as nitric oxide, many electrons are lost in collisions with the molecules forming NO^-. Because the negative ion will not produce an avalanche, and hence a count, the efficiency of the counter is small and not simply related to the ionization yield and window transparency.

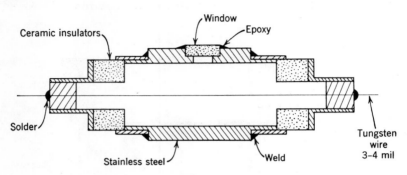

Fig. 7.46 Geiger or photon counter.

Photon counters have generally been used at wavelengths greater than 1040 Å and at wavelengths less than 300 Å [74–81], but not between these wavelength limits. The reason for this is simply the problem of finding a transmitting material sufficiently strong to withstand pressure differentials of several torr. Counters could be used in the range 300–1040 Å if very fine holes or slits were employed to allow the radiation to enter the counter and yet impede the flow of gas from the counter.

The construction of a photon counter is more exacting than the construction of an ion chamber. Excellent details on the preparation and construction of counters have been given by Korff [82]. The most important points to observe are (a) the wire should be accurately parallel to the axis of the cylinder, (b) the wire should be as thin as possible (typically 3 or 4 mil tungsten wire), and (c) there should be no sharp edges or impurities such as dust within the counting volume. The cylinder should be constructed from a metal with a high work function, or else plated with gold. Before the counter is filled, it should be thoroughly outgassed. Furthermore, glowing the wire under vacuum drives off occluded gases and tends to remove microscopic pieces of dust and sharp points on the wire. All traces of alkali metals must be avoided. Figure 7.46 shows a simple construction for a photon counter. The central wire is kept under tension while a heating current is passed through it. The ends of the wire are then soldered in position while the wire is hot. On cooling, the wire is left under tension, thus preventing kinks.

Conventional gas fillings can be used for the spectral range below 300 Å. Mixtures of argon with 10 per cent methane or helium with 4 per cent iso-butane at a total pressure of about 100 torr have performed satisfactorily in the self-quenching mode. For wavelengths above 1040 Å, nitric oxide, iodine, or some polyatomic molecule must be used. Because the absorption

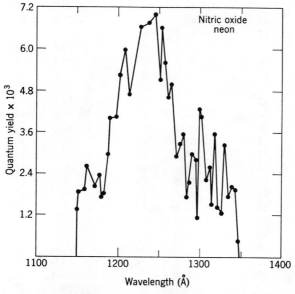

Fig. 7.47 Spectral sensitivity of a nitric oxide quenched counter. Filling: 20 mm NO, 760 mm Ne; window: LiF. (Courtesy T. A. Chubb and H. Friedman [74].)

cross sections are high for these molecules in this region, the radiation is normally totally absorbed at a pressure of 10 to 20 torr, depending on the absorbing path length. At these low pressures, "plateau" characteristics are not too good, and it is customary to increase the pressure with the addition of He or Ne. The spectral sensitivity of a nitric oxide quenched counter, 20 torr NO and 760 torr Ne, as obtained by Chubb and Friedman [74], is shown in Fig. 7.47. The efficiency of the counter at Lyman-α (1216 Å) is about 0.6 per cent. Calibrations at the GCA Corporation using a counter with a pure NO filling of 10 torr gave an efficiency of approximately 0.5 per cent. Completely stable and reproducible results are not easily obtained for prolonged periods using NO. This is possibly due to the fact that NO ions may dissociate into atomic oxygen and atomic nitrogen when they recombine with an electron. The N and O

Table 7.4 Photoionization Yield of Nitric Oxide in the Spectral Range 1062 to 1345 Å[a]

λ (Å)	γ (%)	λ (Å)	γ (%)	λ (Å)	γ (%)
1062.2	67	1163.9	89	1257.1	72
65.1	61	66.1	88	59.9	69
66.3	70	68.8	87	61.7	65
68.3	75	72.1	82	64.2	59
69.8	69	74.4	85	65.7	51
72.7	66	75.9	86	67.2	47
76.7	68	78.3	87	69.0	46
79.8	75	80.5	88	71.4	49
81.6	77	83.1	90	73.0	47
84.5	74	84.4	88	74.1	45
86.8	79	87.9	84	76.7	51
88.6	83	89.4	87	78.0	54
89.9	80	91.7	88	79.6	55
92.7	79	93.3	88	81.0	49
94.7	78	94.6	88	83.2	44
98.1	78	95.7	87	84.3	36
99.5	78	98.2	86	86.6	34
1100.6	78	99.5	83	87.8	37
02.1	84	1201.8	84	89.4	44
04.4	87	04.9	88	90.4	52
05.8	85	06.6	88	93.4	45
07.2	83	08.9	87	95.6	44
10.5	82	11.4	87	97.3	44
11.6	83	12.9	87	99.7	39
15.1	81	15.7	81	1302.4	31
16.5	80	17.4	82	05.0	23
19.1	81	19.0	85	07.4	22
21.3	82	21.1	85	10.1	21
24.0	81	23.5	86	11.0	20
26.0	85	25.6	84	12.9	21
27.3	88	28.3	82	15.4	23.5
31.4	92	30.1	81	16.6	23
32.8	92	32.0	74	19.0	21
35.4	78	34.0	76	21.3	16
37.5	70	35.7	77	23.3	14.5
40.0	79	38.0	75	25.0	13
41.7	77	39.6	80	27.5	14.5
44.4	76	41.4	79	29.1	16
46.0	81	43.5	79	31.0	19
48.6	82	45.9	77	33.6	16.5
50.9	86	47.3	77	35.7	14
53.8	89	50.9	66	38.3	16.5
57.4	87	52.0	67	42.0	6.0
59.8	91	53.6	76	43.3	4.3
61.3	91	55.6	78	45.1	1.3

[a] Courtesy K. Watanabe, private communication.

250

atoms rapidly form NO_2 and N_2 on collision with the neutral molecules by the processes

$$O + NO \rightarrow NO_2 + h\nu \qquad (7.6)$$

and

$$N + NO \rightarrow N_2 + O. \qquad (7.7)$$

The free oxygen released in reaction (7.7) is converted into NO_2 via reaction (7.6). Exposure to intense vacuum uv radiation of longer wavelengths may also cause photodecomposition of NO, thus forming N_2, NO_2, N_2O, and perhaps higher order oxides of nitrogen [83,84]. The formation of different species within the counter tends to alter its characteristics.

The use of nitric oxide is extremely common in both photon counters and ion chambers. For best results, it should be as pure as possible. One simple purification method is to flow commercially pure tank NO through a dry-ice cold trap and then through a zeolite trap at room temperature, collecting the pure NO in a liquid nitrogen trap. The NO can later be expanded into 1 litre flasks. The photoionization yield of NO is tabulated in Table 7.4 for the spectral range 1060 to 1345 Å. The total absorption coefficient at 1216 Å is 64.5 cm^{-1}.

The use of iodine filled counters has been described by Brackmann et al. [81]. With respect to lifetime, iodine appears to be more favorable than nitric oxide. However, since iodine does not exist naturally as a gas, its vapor pressure is highly temperature dependent. Table 7.5 lists the vapor pressure as a function of temperature. At room temperature

Table 7.5   Vapor Pressure of Iodine[a]

Temperature (°K)	Vapor Pressure (torr)
180	1.35×10^{-8}
190	1.32×10^{-7}
200	1.20×10^{-6}
225	8.51×10^{-5}
250	2.51×10^{-3}
275	3.72×10^{-2}
300	3.72×10^{-1}
325	2.63
350	13.2
375	51.3
400	168.0
425	480.0

[a] A. N. Nesmeyanov, *Vapor Pressure of the Elements*, (New York: Academic, 1963) p. 466.

(300°K), the vapor pressure is 372 microns. Brackmann et al. estimate the absorption coefficient of iodine at 1216 Å to be approximately 2800 cm^{-1}; thus at 300°K, 99 per cent of the 1216 Å radiation would be absorbed in a path length of approximately 3.7 cm. They have also estimated the photoionization yield of iodine to be 42 per cent at 1216 Å. Thus at maximum efficiency, an iodine counter would be about a factor of 2 less sensitive than a nitric oxide counter.

Table 7.6 lists window-gas filling combinations suitable for photon counters and ion chambers for a variety of spectral regions in the range 2 to 1715 Å. Figure 7.48 shows the response curves for several of these combinations as measured by Carver and Mitchell [85]. Samson and

Table 7.6   Narrow Band Detectors Derived from Suitable Gas
Filling and Window Combinations

Window	Thickness	Gas	Spectral Response (Å)
Beryllium	0.005 in.	Neon or Argon	2–8
Aluminum	0.00025 in.	Neon	2–6 and 8–16
Mylar	0.00025 in.	Nitrogen or Helium	2–15 and 44–60
Cellulose nitrate or Zapon	1000 Å[a]	Argon	<300
Windowless		Helium	<504.3
		Neon	<574.9
		Argon	<786.7
		Krypton	<885.6
		Xenon	<1022.1
Lithium fluoride	1 mm	Ethyl chloride	1040–1130
		Ethyl bromide	1040–1200
		Carbon disulfide	1040–1240
		Acetone	1040–1290
		Nitric oxide	1040–1340
Magnesium fluoride	1 mm	Carbon disulfide	1120–1240
		Nitric oxide	1120–1340
Calcium fluoride	1 mm	Acetone	1220–1290
		Ethyl iodide	1220–1330
		Nitric oxide	1220–1340
		Benzene	1220–1340
Barium fluoride	1 mm	Toluene	1340–1410
		p-Xylene	1340–1470
Sapphire	1 mm	p-Xylene	1410–1470
		Mesitylene	1410–1480
Fused quartz	1 mm	tri-n-Propyl amine	1600–1715

[a] Cannot withstand a pressure differential of 1 atmosphere.

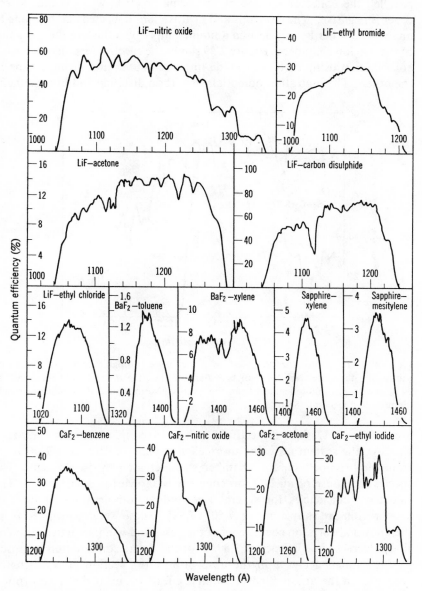

Fig. 7.48 Response curves for ion chambers using various windows and gases in the wavelength range 1000 to 1350 Å (courtesy J. H. Carver and P. Mitchell [85]).

253

Golomb [86] have shown the effect of connecting two ion chambers in parallel, the collector voltage of one being positive, whereas that of the other is negative. The electrons and positive ions collected can be made to annul each other by any desired amount simply by adjusting the gas-gains of the two ion chambers. Figure 7.49 shows the spectral response of such a combination using carbon disulfide in one chamber and bromoethane in the other. The ionization potentials of carbon disulfide and bromoethane

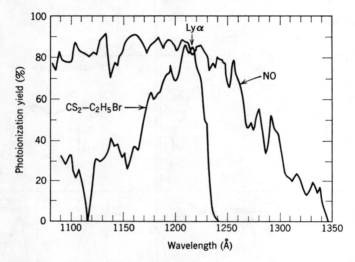

Fig. 7.49 The photoionization yield of nitric oxide compared to that obtained by subtracting the "effective" yield of bromoethane (i.e., the normal yield increased by 14% due to gas multiplication) from the photoionization yield of carbon disulphide (Samson and Golomb [86]).

are 1240 and 1210 Å, respectively. Used in conjunction with a MgF_2 window, the system provides an excellent Lyman-α detector. The signal from the bromoethane ion chamber was increased by 14 per cent. The spectral response of an NO ion chamber is shown for comparison. The calculated efficiencies for several gas-window combinations in the soft x-ray region are shown in Fig. 7.50 [87,88]. These were determined from the known absorption coefficients of the gases and window materials. The sharp decrease in efficiency at 3.87 Å for the Be-Ar combination is caused by the sudden decrease in the absorption cross section at wavelengths greater than the argon K edge. Similar effects occur at the K-absorption edges of Al (8 Å) and C (43.6 Å).

Finally, Table 7.7 tabulates the ionization potentials for many gases and vapors which may be suitable as gas fillings. They have been selected to cover as wide a spectral range as possible. Extensive tables of ionization

Table 7.7 Ionization Potentials of Some Gases and Vapors[a]

Substance		Ionization Potential (eV)	(Å)
Helium	He	24.58	504.3
Neon	Ne	21.56	574.9
Argon	Ar	15.76	786.7
Krypton	Kr	14.00	885.6
Methane	CH_4	12.98	955
Nitrous Oxide	N_2O	12.90	961
Xenon	Xe	12.13	1022.1
Chlorine	Cl_2	11.48	1080
Methyl chloride	CH_3Cl	11.28	1099
Ethyl chloride	C_2H_5Cl	10.98	1129
Isobutane	C_4H_{10}	10.57	1173
Hydrogen sulfide	H_2S	10.46	1185
Ethyl bromide	C_2H_5Br	10.29	1205
Boron trifluoride	BF_3	10.25	1210
Ethyl acetate	$CH_3COOC_2H_5$	10.11	1226
Carbon disulfide	CS_2	10.08	1230
Nitrogen dioxide	NO_2	9.78	1268
Acetone	$(CH_3)_2CO$	9.69	1279
Iodine	I_2	9.28	1336
Nitric oxide	NO	9.25	1340
Benzene	C_6H_6	9.25	1340
Methyl amine	CH_3NH_2	8.97	1382
Ethyl amine	$C_2H_5NH_2$	8.86	1400
t-Butyl amine	$(CH_3)_3CNH_2$	8.64	1435
Di-methyl amine	$(CH_3)_2NH$	8.24	1505
Di-ethyl amine	$(C_2H_5)_2NH$	8.01	1548
Tri-methyl amine	$(CH_3)_3N$	7.82	1585
Tri-ethyl amine	$(C_2H_5)_3N$	7.50	1653
Tri-n-propyl amine	$(C_3H_7)_3N$	7.23	1715

[a] The ionization potentials for all gases and vapors with the exception of BF_3 and the rare gases were obtained from K. Watanabe, T. Nakayama, and J. Mottl, *J. Quant. Spectrosc. Radiat.* Transfer **2**, 369 (1962); BF_3 from B. Kaufman, *Phys. Rev.* **78**, 332 (1950); Rare gases from C. E. Moore, Natl. Bur. Std. (U.S.), Circ. **476**, Vol. 1 (1949), Vol. 2 (1952), Vol. 3 (1958).

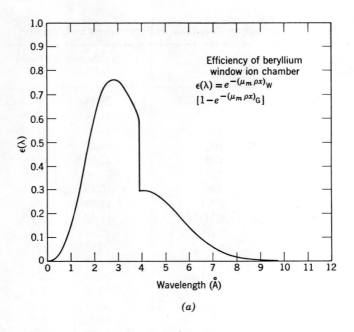

(a)

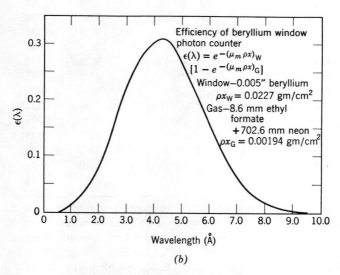

(b)

Fig. 7.50 Calculated efficiencies for photon counters with different gas-window combinations (courtesy R. W. Kreplin et al. [87, 88]).

256

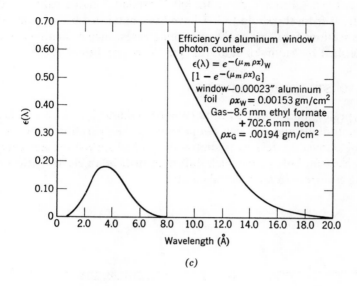

(c)

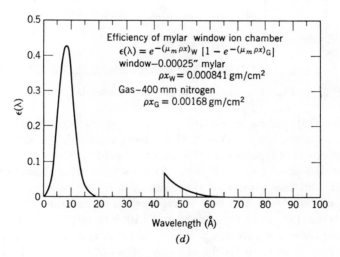

(d)

Fig. 7.50 (*continued*)

257

potentials have been published by Watanabe et al. [89] and by Higasi et al. [90]. The photoionization cross sections of the rare gases are given in Chapter 8, Figs. 8.4 to 8.7, down to 300 Å, while tabulated values of the cross sections from threshold to 1 Å have been published by Samson [91]. The cross sections for some of the quenching gases, such as methane, have been published by Lukirskii et al. [92] for the region below 200 Å.

7.5 SOLID STATE PHOTODIODES

Two types of solid state detectors have been studied, a silicon surface-barrier photodiode and a silicon p-n junction. The operating principle of these semiconducting devices is simply that the absorbed radiant energy produces electron-hole pairs which diffuse or drift in an electric field thus producing a current in an external circuit.

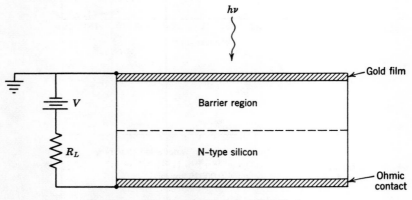

Fig. 7.51   Silicon surface-barrier photodiode.

Tuzzolino [93] studied the efficiency of a silicon surface-barrier photodiode in the spectral range 584 to 2537 Å. The photodiode was fabricated from single crystal n-type silicon with a layer of gold approximately 100 Å thick deposited on one face. Figure 7.51 shows the arrangement of the photodiode and the associated electrical circuit. Because of the low resistivity of the diode, the bias voltage produces currents of 10^{-8} to 10^{-7} A, whereas the photocurrents are in the 10^{-10} to 10^{-12} A range. Thus dc recording techniques are not practical, and it is necessary to chop the incident radiation and amplify the ac signal. The output signal is found to be independent of the magnitude of the applied bias voltage. The efficiency of this photodiode is shown in Fig. 7.52 as a function of wavelength, where the efficiency is defined as the number of electrons flowing in the external load resistor (R_L) per photon incident on the photodiode.

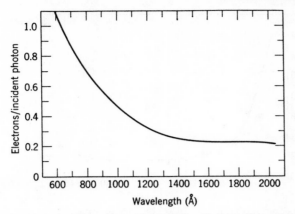

Fig. 7.52 Spectral sensitivity for a silicon surface-barrier photodiode with an area of 2.75 cm² and a nominal gold film thickness of 100 Å (courtesy A. J. Tuzzolino [93]).

Morse et al. [94] have used a silicon *p-n* junction photodiode as a detector for the wavelength range 584 to 1600 Å. Their arrangement is shown in Fig. 7.53. Using *p*-type silicon, they introduced phosphorus as an *n*-type impurity. The silicon dioxide that forms on the surface is etched to expose the photosensitive area. To minimize surface leakage currents, a guard ring is used. In addition to producing an internal photo-current (I_p), the incident radiation also ejects photoelectrons. The photo-diode shown in Fig. 7.53 suppresses these electrons by the use of a retarding

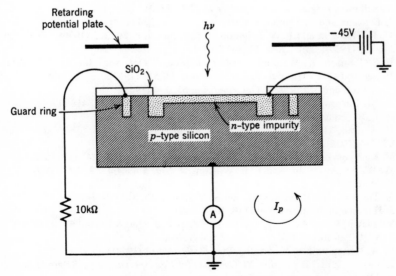

Fig. 7.53 Silicon *p-n* junction photodiode.

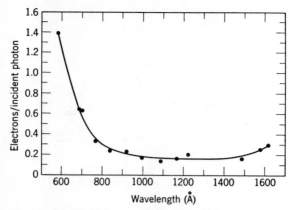

Fig. 7.54 Spectral sensitivity for a silicon *p-n* junction photodiode (courtesy A. L. Morse et al. [94]).

potential plate placed above the sensitive area. The efficiency of the silicon *p-n* junction photodiode is shown in Fig. 7.54.

Both photodiodes described above show promise as lightweight miniaturized detectors, especially for the shorter wavelengths if the efficiency continues to increase linearly with the photon energy.

REFERENCES

[1] V. Schumann, *Wien. Akad. Anzeiger* **23**, 230 (1892).

[2] J. Duclaux and P. Jeantet, *J. Phys. et Radium* **2**, 154 (1921).

[3] G. R. Harrison and P. A. Leighton, *J. Opt. Soc. Am.* **20**, 313 (1930); *Phys. Rev.* **34**, 779 (1930); *Phys. Rev.* **38**, 899 (1931).

[4] F. S. Johnson, K. Watanabe, and R. Tousey, *J. Opt. Soc. Am.* **41**, 702 (1951).

[5] P. Lee and G. L. Weissler, *J. Opt. Soc. Am.* **43**, 512 (1953).

[6] R. Allison and J. Burns, *J. Opt. Soc. Am.* **55**, 574 (1965).

[7] K. M. Burrows, J. C. Kelly, and D. E. Ellis, *Aust. J. Phys.* **17**, 418 (1964).

[8] A. L. Shoen and E. S. Hodge, *J. Opt. Soc. Am.* **40**, 23 (1950).

[9] R. Audran, "Intern. Konf. wiss. Phot. Köhn.," p. 279 (1956); *Sci. Indust. Phot.* **27**, 434 (1956).

[10] V. K. Zworykin, G. A. Morton, and L. Malter, *Proc. I.R.E.* **24**, 351 (1936).

[11] W. W. Parkinson, Jr., and F. E. Williams, *J. Opt. Soc. Am.* **39**, 705 (1949).

[12] K. Watanabe and E. C. Y. Inn, *J. Opt. Soc. Am.* **43**, 32 (1953).

[13] G. Déjardin and R. Schwégler, *Revue D'Optique* **13**, 313 (1934).

[14] J. Hammann, *Z. Angew. Phys.* **10**, 187 (1958).

[15] W. Slavin, R. W. Mooney, and D. T. Palumbo, *J. Opt. Soc. Am.* **51**, 93 (1961).

[16] J. A. R. Samson, *J. Opt. Soc. Am.* **54**, 6 (1964).

[17] R. A. Knapp and A. M. Smith, *Appl. Optics* **3**, 637 (1964).

[18] J. Vasseur and M. Cantin, *XI Colloque Int'l de Spectroscopie*, Belgrade (1963), p. 491.

[19] E. C. Bruner, Jr., Thesis, University of Colorado, Colorado (1964).

[20] R. Allison, J. Burns, and A. J. Tuzzolino, *J. Opt. Soc. Am.* **54,** 1381 (1964).

[21] M. Seya and F. Masuda, *Sci. Light* **12,** 9 (1963).

[22] R. Allison, J. Burns, and A. J. Tuzzolino, *J. Opt. Soc. Am.* **54,** 747 (1964).

[23] K. J. Nygaard, *Brit. J. Appl. Phys.* **15,** 597 (1964).

[24] J. C. Lemonnier, M. Priol, A. Quemerais, and S. Robin, *J. Physique* **25,** 79A (1964).

[25] D. H. Thurnau, *J. Opt. Soc. Am.* **46,** 346 (1956).

[26] K. J. Nygaard and R. S. Sigmond, results quoted by J. M. Breare and A. von Engel, *Proc. Roy. Soc. A.* **282,** 394 (1964).

[27] K. J. Nygaard, private communication (March 1965).

[28] G. K. Herb and W. J. Van Sciver, *Rev. Sci. Instr.* **36,** 1650 (1965).

[29] J. A. R. Samson, unpublished data (1965).

[30] F. J. Studer, *J. Opt. Soc. Am.* **55,** 615A (1965), and private communication (November 1965).

[31] K. J. Nygaard, *J. Opt. Soc. Am.* **55,** 944 (1965).

[32] H. Inokuchi, Y. Harada, and T. Kondow, *J. Opt. Soc. Am.* **54,** 842 (1964).

[33] N. Kristianpoller, *J. Opt. Soc. Am.* **54,** 1285 (1964).

[34] J. A. R. Samson and R. B. Cairns, *Appl. Optics* **4,** 915 (1965).

[35] R. Lincke and G. Palumbo, *Appl. Optics* **4,** 1677 (1965).

[36] M. Brunet, M. Cantin, C. Julliot, and J. Vasseur, *J. de Physique* **24,** 53A (1963).

[37] N. Kristianpoller and D. Dutton, *Appl. Optics* **3,** 287 (1964).

[38] R. Conklin, *J. Opt. Soc. Am.* **49,** 669 (1959).

[39] L. Dunkelman, W. B. Fowler, and J. Hennes, *Appl. Optics* **1,** 695 (1962).

[40] P. H. Metzger, *J. Phys. Chem. Solids* **26,** 1879 (1965).

[41] R. B. Cairns and J. A. R. Samson, *J. Opt. Soc. Am.* **56,** 1568 (1966).

[42] W. C. Walker, N. Wainfan, and G. L. Weissler, *J. Appl. Phys.* **26,** 1366 (1955).

[43] K. Watanabe, University of Hawaii, unpublished data (1964).

[44] J. A. R. Samson and R. B. Cairns, *Rev. Sci. Instr.* **36,** 19 (1965).

[45] J. S. Allen, *Rev. Sci. Instr.* **12,** 484 (1941). *See also Phys. Rev.* **55,** 966 (1939); *Rev. Sci. Instr.* **18,** 739 (1947); and *Proceedings of the I.R.E.* p. 346, April 1950.

[46] A. P. Lukirskii, M. A. Rumsh, and L. A. Smirnov, *Optics and Spectroscopy* **9,** 265 (1960).

[47] A. P. Lukirskii, M. A. Rumsh, and I. A. Karpovich, *Optics and Spectroscopy* **9,** 343 (1960).

[48] M. A. Rumsh, A. P. Lukirskii, and V. N. Shchemelev, *Soviet Physics* Doklady **135,** 1231 (1960).

[49] A. P. Lukirskii, E. P. Savinov, I. A. Brytov, and Yu. F. Shepelev, *USSR Acad. Sci. Bulletin Phys.* Ser. **28,** 774 (1964).

[50] E. P. Savinov, A. P. Lukirskii, and Yu. F. Shepelev, *Fiz. trend. Tela* **6,** 3279 (1964). English trans., *Soviet Phys.-Solid State* **6,** 2624 (1965).

[51] J. L. Stanford, R. N. Hamm, and E. T. Arakawa, *J. Opt. Soc. Am.* **56,** 124 (1966).

[52] H. E. Hinteregger, in *Space Astrophysics,* ed. W. Liller (McGraw-Hill, New York, 1961).

[53] L. Heroux and H. E. Hinteregger, *Rev. Sci. Instr.* **31,** 280 (1960).

[54] G. W. Goodrich and W. C. Wiley, *Rev. Sci. Instr.* **32,** 846 (1961).

[55] G. W. Goodrich and W. C. Wiley, *Rev. Sci. Instr.* **33,** 761 (1962).

[56] W. R. Hunter, *Space Research III,* ed. W. Priester (North-Holland, Amsterdam, 1963) p. 1187.

[57] J. Adams and B. W. Manley, *Electronic Engineering* **37,** 180 (1965).

[58] R. Lincke and T. D. Wilkerson, *Rev. Sci. Instr.* **33**, 911 (1962).

[59] N. R. Daly, *Rev. Sci. Instr.* **31**, 264 and 720 (1960).

[60] E. Schönheit, *A. Naturforschg.* **15a**, 839 (1960).

[61] J. A. R. Samson and R. B. Cairns, *Rev. Sci. Instr.* **37**, 338 (1966).

[62] L. Heroux, J. E. Manson, and H. E. Hinteregger, *J. Opt. Soc. Am.* **55**, 103 (1965).

[63] G. L. Locher, *Phys. Rev.* **42**, 525 (1932).

[64] O. S. Duffendack and W. E. Morris, *J. Opt. Soc. Am.* **32**, 8 (1942).

[65] G. E. Mandeville and M. V. Scherb, *Nucleonics* **7**, 34 (1950).

[66] D. W. Turner, *Nature*, **178**, 1022 (1957).

[67] D. W. Turner, *J. Chem. Soc.* p. 4555 (1957).

[68] H. R. Phillip and E. A. Taft, *J. Phys. Chem. Solids* **1**, 159 (1956).

[69] W. Pong, University of Hawaii, unpublished data (1964).

[70] M. Rome, *IEEE Transactions on Nuclear Science*, June 1964.

[71] A. I. Efremov, A. L. Podmoshenskii, M. A. Ivanov, V. N. Nikeforov, and O. N. Efimov, *Isk. Sputniki Zemli* **10**, 48 (1961). English trans., *Planet. Space Sci.* **9**, 987 (1962).

[72] G. Skorinko, D. D. Doughty, and W. A. Feibelman, *Appl. Optics* **1**, 717 (1962).

[73] R. E. Benn, W. S. Foote, and C. T. Chase, *J. Opt. Soc. Am.* **39**, 529 (1949).

[74] T. A. Chubb and H. Friedman, *Rev. Sci. Instr.* **26**, 493 (1955).

[75] H. Friedman, S. W. Lichtman, and E. T. Byram, *Phys. Rev.* **83**, 1025 (1951).

[76] A. P. Lukirskii, M. A. Rumsh, and L. A. Smirnov, *Optics and Spectroscopy* **9**, 262 (1960).

[77] J. E. Holliday, *Rev. Sci. Instr.* **31**, 891 (1960).

[78] J. L. Rogers and F. C. Chalklin, *Proc. Phys. Soc.* **B67**, 348 (1954).

[79] D. L. Ederer and D. H. Tomboulian, *Appl. Optics* **3**, 1073 (1964).

[80] A. J. Caruso and W. M. Neupert, *Appl. Optics* **4**, 247 (1965).

[81] R. T. Brackmann, W. L. Fite, and K. E. Hagen, *Rev. Sci. Instr.* **29**, 125 (1958).

[82] S. A. Korff, *Electron and Nuclear Counters* (Van Nostrand, New York, 1957), 2nd ed.

[83] P. J. Flory and H. L. Johnson, *J. Chem. Phys.* **14**, 212 (1946).

[84] J. J. McGee and J. Heicklen, *J. Chem. Phys.* **41**, 2974 (1964).

[85] J. H. Carver and P. Mitchell, *J. Sci. Instr.* **41**, 555 (1964).

[86] J. A. R. Samson and D. Golomb, *Rev. Sci. Instr.* **34**, 441 (1963).

[87] R. W. Kreplin, *Annales de Geophysique* **17**, 151 (1961).

[88] R. W. Kreplin, T. A. Chubb, and H. Friedman, *J. Geophys. Research* **67**, 2231 (1962).

[89] K. Watanabe, T. Nakayama, and J. Mottl, *J. Quant. Spectrosc. Radiat. Transfer* **2**, 369 (1962).

[90] K. Higasi, I. Omura, and T. Tsuchiya, *Tables of Ionization Potentials of Molecules and Radicals* (Research Inst. of Appl. Elect., Hokkaido University, Japan, 1956).

[91] J. A. R. Samson, in *Advances in Atomic and Molecular Physics*, eds. D. R. Bates and I. Esterman (Academic, New York, 1966), Vol. 2, p. 177.

[92] A. P. Lukirskii, I. A. Brytov, and T. M. Zimkina, *Optics and Spectroscopy* **17**, 234 (1964).

[93] A. J. Tuzzolino, *Rev. Sci. Instr.* **35**, 1332 (1964); *Phys. Rev.* **134**, A205 (1964).

[94] A. L. Morse, W. F. Crevier, D. B. Medved, and G. L. Weissler, University of Southern California, Los Angeles, unpublished data (1965).

8

Absolute Intensity Measurements

8.1 INTRODUCTION

To measure the absolute intensity of radiation of any wavelength, a detector must be used whose response is known in absolute units. The detector must either be calibrated against a standard source of known intensity or it must be an *absolute detector*, whose response to the intensity of radiation can be predicted.

The calorimetric measurement of the heat produced by complete absorption of the incident radiation would be a measure of the absolute intensity of the radiation provided the interaction of the radiation with the calorimeter was completely converted into heat. The calorimeter would then be an absolute detector. Unfortunately, such a detector is very insensitive to the source intensities commonly available and does not constitute a practical detector. However, thermal detectors such as the thermocouple which transform heat energy into electrical energy are much more sensitive than the calorimeter. The thermocouple has long been used as a primary standard, but it is not an absolute standard; it must be calibrated against a standard source of radiation of known intensity.

The reason that the thermocouple has been chosen as a primary standard is that its response, in μ volts per μ watt, is independent of wavelength from the infrared to the soft x-rays. It is this "flatness of response" that makes it possible to use a thermocouple to measure absolute intensities in the vacuum uv when the thermocouple has been calibrated with visible radiation. Unfortunately, in the vacuum ultraviolet region of the spectrum, the flux density emitted by many light sources is far weaker than in the visible, especially after transmission through a monochromator where the initial radiant intensity is greatly attenuated by the poor reflectance of gratings (see Chapter 2). This necessitates the use of wide slits in the monochromator to obtain sufficient light intensity for a measurable signal from the thermocouple. Wavelength resolution is sacrificed for this increased intensity; however, a more sensitive standard, typically a sodium salicylate coated photomultiplier, can then be calibrated against the thermocouple. With this more sensitive secondary standard, the monochromator slits can be narrowed and a higher wavelength resolution achieved. Using the above technique, Watanabe et al. [1] measured the photoelectric yield of nitric oxide, among many other gases. Of particular interest is the yield of NO at Lyman-α (1215.7 Å), which has recently been remeasured by Watanabe and found to lie between 80 and 85 per cent. This value for the yield of NO has been used as a standard to determine the absolute intensity of the hydrogen Lyman-α line emitted from the sun using nitric oxide ionization chambers in rockets and satellites. However, the many steps involved in reaching the photoelectric yield of NO increases the chance for error in the value of the yield.

Because of the lack of absolute detectors, intensity measurements have in the past been based on absolute sources of radiation such as the ideal black body furnace. Secondary standard sources are then constructed and calibrated against the primary black body source. These secondary standard sources are typically specially constructed tungsten filament lamps produced by the National Bureau of Standards. The emission of these lamps is confined to wavelengths greater than 2700 Å. To measure absolute intensities at shorter wavelengths, it is necessary to select a stable detector whose response is constant with respect either to wavelength or frequency since the detector must be calibrated at the longer wavelengths; hence the importance of thermocouples. The product of the voltage generated by the thermocouple and the wavelength of the incident radiation is proportional to the number of photons per second incident on the thermocouple. Recently, however, the calibration of thermocouples directly with vacuum uv radiation has become possible using special constructed plasma arcs that approximate black body radiators at selected wavelengths characteristic of the gas used in the arc. Perhaps of more importance is the application of synchrotron radiation to the calibration of detectors. The synchrotron radiation is continuous from the far infrared to the x-ray region, and its intensity can be precisely calculated for any wavelength interval. These absolute sources are described in the following sections.

An absolute detector can be constructed utilizing the principle of photoionization of a rare gas [2]. Over a certain wavelength range, every photon absorbed by the gas produces an ion. Thus if total absorption of the light occurs, the number of photons incident on the gas is simply equal to the ion current produced. The long wavelength limit of such an absolute detector is 1022 Å, the ionization potential of xenon. The short wavelength limit using helium is about 250 Å. For shorter wavelengths, the ejected photoelectrons have sufficient energy to cause secondary ionization. When this is the case, a carefully constructed Geiger counter can be made into an absolute detector [3,4].

The photoionization technique is described in detail in the next section followed by a discussion of thermocouples and other thermal detectors. The remaining sections describe absolute sources and some rather interesting and independent techniques for absolute intensity measurements.

8.2 PHOTOIONIZATION DETECTORS

By far the most accurate and reproducible method of measuring absolute intensities is that utilizing photoionization of a suitable gas. Actually, any gas would suffice provided its photoionization yield is known. The photoionization yield of a gas is defined as the number of ions produced per photon absorbed by the gas. All polyatomic gases have, in general, yields which are equal to or less than 100 per cent. One reason for yields less than 100 per cent is that the photon energy may be absorbed in dissociating the molecule and leaving some or all of the constituent atoms in excited states. Atoms, on the other hand, have yields of 100 per cent. Possible exceptions may occur in regions of discrete structure superimposed on the ionization continuum. However, the interaction between the continuum and the discrete levels generally results in a radiationless transition into the continuum with the release of a photoelectron. This process is called *autoionization*. The ratio of the number of photoelectrons emitted to the total number of atoms excited into an autoionized state is equal to $\alpha/(\beta + \alpha)$, where α and β are, respectively, the probabilities for radiationless and radiative transitions. The probability for a given transition (single electron transition) is equal to the inverse of the mean lifetime of that state. Therefore, since the lifetime of radiationless transitions into the ionization continuum is commonly of the order of 10^{-13} to 10^{-15} secs compared to 10^{-8} secs for a radiative transition, then $\alpha/(\beta + \alpha)$ is essentially unity. Thus, the photoionization yield of an atom is expected to be unity in an autoionized level.

The rare gases (helium, neon, argon, krypton and xenon) are our only source of permanent atomic gases. The photoionization yields of these gases have been measured relative to one another, even in regions of auto-ionization, and shown to be 100 per cent [2,5]. Thus, if an ionization chamber is constructed and filled with the appropriate rare gas until all of the incident radiation is absorbed, then the number of photons per sec incident on the gas is simply equal to the ion current produced. In the case of a photon counter, the number of photons per sec will be equal to the number of pulses per sec produced by the counter. To achieve this one-to-one correspondence between the number of photons incident and the number of ions or pulses produced, the ion chambers and photon counters must be carefully constructed. The following sections outline the principles involved. The ion chamber can be used as an absolute detector of radiation from 250 to 1022 Å, while the photon counter can be used from 2 to 300 Å.

Ion Chambers

Several designs for a standard ion chamber are possible. Three designs are described below in detail. These ion chambers can all be used to measure absolute intensities, even when the incident radiation is not completely absorbed by the gas. Two of the ion chambers are also suitable when complete absorption takes place. Under certain circumstances, one design may be more suitable than another.

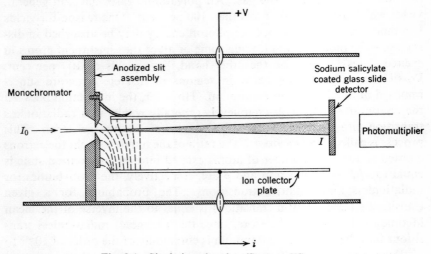

Fig. 8.1   Single ion chamber (Samson [2]).

Single Ion Chamber.   Referring to Fig. 8.1 for an analysis of the single ion chamber, we define I_0 as the light intensity at the exit slit of the monochromator and I as the transmitted intensity at the end of the ion chamber. The ion current i is expressed in amperes. Then,

$$\gamma = \frac{\text{ions formed/sec}}{\text{photons absorbed/sec}}$$

$$= \frac{i/e}{I_0 - I}$$

$$= \frac{i/e}{I_0(1 - I/I_0)}.$$

Therefore

$$I_0\gamma = \frac{i/e}{(1 - I/I_0)},$$ (8.1)

where e is the electronic charge and is equal to 1.6×10^{-19} coulomb.

The ratio I/I_0 is measured by the detector lying exactly at the end of the ion chamber. This ratio is independent of absolute intensities, and any detector which has a linear response with respect to intensity may be used. This method requires that all the ions formed from the exit slit to the detector be collected and counted. To achieve this, it is necessary to connect the exit slit electrically to the positive repeller plate. The ion chamber will then have a field distribution as shown in Fig. 8.1, and all ions formed within the ion chamber system will be collected. The major advantage of this system is that no measurement of an absorption coefficient is made which must obey Beer's law. Actually, I_0 is independent of the pressure used, and in the limit when $I/I_0 \to 0$, (8.1) becomes

$$I_0\gamma = \frac{i}{e}.$$ (8.2)

Double Ion Chamber. With the aid of Fig. 8.2, an analysis of the double ion chamber is given. Denote I_0 as the incident light intensity to be measured; then with a gas in the cell at some suitable pressure, say P, denote I_1 and I'_1 as the intensities entering and leaving plate 1, and denote I_2 and I'_2 as the intensities entering and leaving plate 2. The three small plates alternating with the ion collection plates are guard rings to provide a uniform field between the parallel plates, and all are at ground potential. The repeller plate is held a few volts positive to drive the ions to the

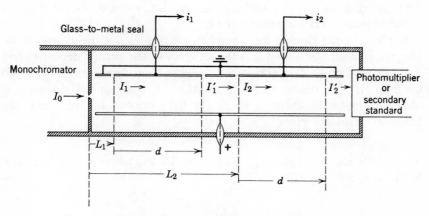

Fig. 8.2   Double ion chamber (Samson [2]).

collector plates 1 and 2. Now at plate 1, the total number of ions produced per second is equal to i_1/e, where i_1 is the electric current in amperes, and e is the electronic charge measured in coulombs. The total number of photons absorbed per second is equal to $(I_1 - I'_1)$. Using Lambert's law, this becomes

$$I_1 - I'_1 = I_0 e^{-\mu L_1}(1 - e^{-\mu d}),$$

where μ is the absorption coefficient of the gas measured at a pressure P; L_1 and d are the dimensions shown in Fig. 8.2. From the definition of the photoionization yield, we get

$$\gamma = \frac{i_1/e}{I_0 e^{-\mu L_1}(1 - e^{-\mu d})},$$

or in terms of the experimentally determined quantity,

$$I_0\gamma = \frac{i_1/e}{e^{-\mu L_1}(1 - e^{-\mu d})}. \tag{8.3}$$

Similarly, for plate 2,

$$I_0\gamma = \frac{i_2/e}{e^{-\mu L_2}(1 - e^{-\mu d})}. \tag{8.4}$$

From the ratio of (8.3) and (8.4), and solving for μ, we get

$$\mu = \frac{\ln(i_1/i_2)}{L_2 - L_1}. \tag{8.5}$$

Substituting μ into (8.3), and assuming $\gamma = 1$, I_0 can be found from measurements of L_1, L_2, d, and the ion currents i_1 and i_2.

The same procedure is used to measure γ in other gases. When I_0 is known, (8.3) is again used, and the value of the photoionization yield is determined. It should be noted that the absolute value of the gas pressure is not required in these measurements.

Implicit in the double ion chamber method is the use of measured absorption coefficients, and thus the accuracy of the method depends upon the measured absorption coefficients obeying Beer's law, which states that the amount of light absorbed is proportional to the number of absorbing atoms or molecules through which the light passes. Incorporated into Lambert's law, this gives

$$I = I_0 e^{-kx},$$

where k is the absorption coefficient at S.T.P., whereas x is the path length reduced to S.T.P. and is given by

$$x = L\frac{P}{760}\frac{273}{T}.$$

Thus to obey Beer's law, k must be independent of both pressure P and path length L. This will not be the case in regions of discrete absorption when the absorption lines are narrower than the band pass of the photo-ionizing radiation. Thus in using the double ion chamber to measure absolute intensities or photoionization yields, it is better to avoid regions of sharp discrete structure.

The major advantage of the double ion chamber lies in the fact that all the variables, namely, the two ion currents and, if desired, a secondary standard output, can be measured simultaneously, thereby eliminating any discrepancies due to light source fluctuations.

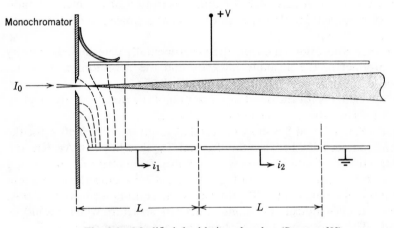

Fig. 8.3 Modified double ion chamber (Samson [2]).

Modified Double Ion Chamber. As shown in Fig. 8.3, the modified double ion chamber is obtained by letting L_1 and L_2 in Fig. 8.2 go respectively to zero and d. In this case, (8.3) becomes

$$I_0\gamma = \frac{i_1/e}{1 - e^{-\mu d}},$$

or

$$I_0\gamma = \frac{(i_1)^2/e}{i_1 - i_2},\tag{8.6}$$

since

$$\mu = \frac{1}{d}\ln\left(\frac{i_1}{i_2}\right).\tag{8.7}$$

As in the case of the single ion chamber, the exit slit of the monochromator must be at the same positive potential as the repeller plate. When the

Table 8.1   Ionization Potentials of the Rare Gases

Gas	Ionization Potential (Å)	
	$^2P_{3/2}$	$^2P_{1/2}$
He	504.26	
Ne	574.93	572.37
Ar	786.72	777.96
Kr	885.62	845.42
Xe	1022.14	922.75

gas pressure is sufficiently high for total absorption of the incident radiation, the modified double ion chamber acts as a single ion chamber with $I_0\gamma = i_1/e$.

With each ionization chamber, the experimentally determined quantity is the product of the absolute intensity and the photoionization yield. Knowledge of one, therefore, provides the other.

The wavelengths for the onset of ionization of the rare gases are given in Table 8.1. The table also includes the $^2P_{1/2}$ onset.

Figures 8.4, 8.5, and 8.6 show the total absorption coefficients k for the autoionized lines lying between the $^2P_{3/2}$ and $^2P_{1/2}$ thresholds of Ar, Kr, and Xe, respectively. These k-values were obtained from reference [6] and are believed to be about 15 per cent high. Figure 8.7 provides the k-values in the continuum starting at the $^2P_{1/2}$ threshold and continuing down to 280 Å. The vertical lines indicate positions of discrete structure which should be avoided. The precise position of these lines are given in reference 7.

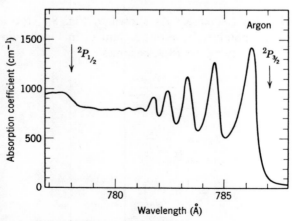

Fig. 8.4 Absorption coefficients of argon in the region of autoionization (courtesy R. E. Huffman et al. [6]).

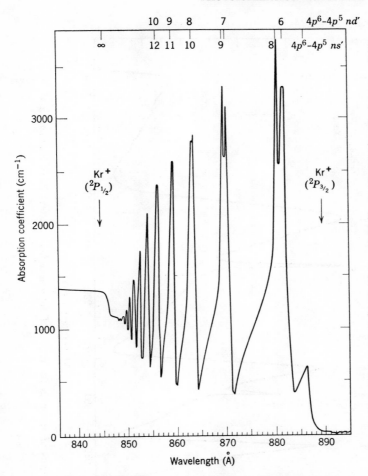

Fig. 8.5 Absorption coefficients of krypton in the region of autoionization (courtesy R. E. Huffman et al. [6]).

The use of photoionization techniques for absolute intensity measurements requires some care. To begin with, because there are no windows available in this spectral region we must take care that the main vacuum system of the monochromator is maintained at a high vacuum to ensure that the photon intensity at the exit slit of the monochromator is the same before and after the calibrating gas is allowed to flow into the ion chamber. Another problem associated with a flow system is to assure that the gas is uniformly at the same pressure throughout the ion chamber. Any pressure gradients would be detrimental when using the double ion chamber. The first problem is minimized by using a long ion chamber, say greater than

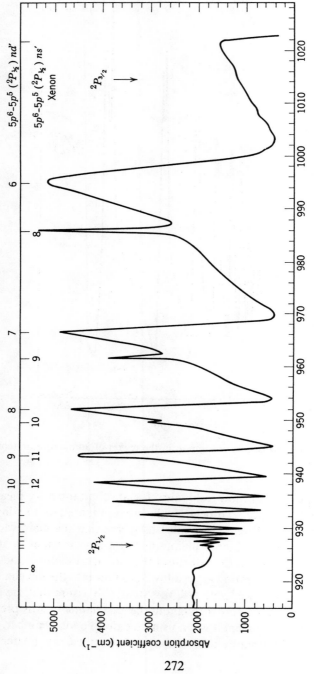

Fig. 8.6 Absorption coefficients of xenon in the region of autoionization (courtesy R. E. Huffman et al. [6]).

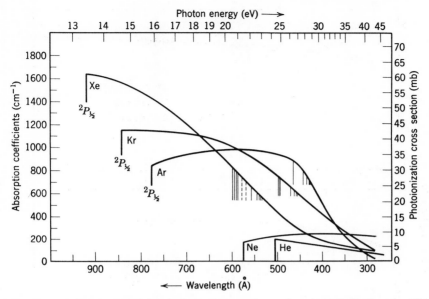

Fig. 8.7 Absorption coefficients of the rare gases as a function of wavelength. The vertical lines indicate positions of discrete structure (Samson [7]).

30 cm, so that less gas pressure is required to absorb all of the incident radiation. The second problem is reduced by using a large diameter ion chamber with a very small entrance slit area. Typical dimensions which have been used successfully are chamber, 7 cm diameter and slits, 0.05 mm wide and 1 cm long.

Probably one of the most important precautions to take is in using the proper collector voltage on the ion plates; that is, we must operate in the plateau region of the ions versus voltage curve. In fact, we must ascertain that there actually is a plateau for the particular ion chamber, calibrating gas, and wavelength used. For example, if xenon were ionized by 461.5 Å (26.86 eV), an electron could be ejected with an energy E given by

$$E = h\nu - \text{I(Xe)}, \tag{8.8}$$

where I(Xe) is the ionization potential of xenon and is equal to 12.13 eV, $h\nu$ is the energy of the incident photon, which in this example is 26.86 eV. Substituting these values into (8.8) gives $E = 14.73$ eV. Thus, the electron is emitted with sufficient energy to cause secondary ionization. With the addition of a collector voltage, this energy is increased. It is, therefore, impossible to achieve a plateau with xenon at 461.5 Å and at a pressure which would give a measurable ion current. The top curve in Fig. 8.8 illustrates this point for a particular ion chamber.

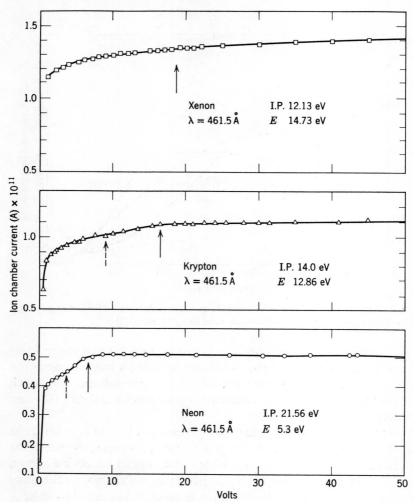

Fig. 8.8 Ion chamber currents as a function of voltage for Xe, Kr, and Ne at 461.5 Å. In each case, the dashed vertical arrow indicates the voltage necessary to commence electron retardation whereas the solid arrow indicates the voltage necessary to complete electron retardation. E represents the energy of the ejected electron due to radiation of wavelength λ (Samson [2]).

As seen in Fig. 8.9, the ionizing radiation is made to pass as close to the ion repeller plate as possible without actually striking it. The reason for this is to provide a maximum retarding potential for electrons travelling towards the collector plate while giving a minimum acceleration to electrons travelling towards the $+V$ plate, thereby decreasing the chance of providing the electrons with sufficient energy to cause secondary ionization.

A further competing process is the collection of electrons as well as ions. This can occur if an insufficiently high electron retarding potential is used. Reference to Fig. 8.9 illustrates possible electron trajectories. These electrons all have the same energy, but those travelling at an angle to the field lines can be retarded at lower voltages than those travelling parallel to, but opposing, the field lines. That is, the important quantity is the component of the electron velocity normal to the collector plate. It has been shown by Sommerfeld [8] that the probability for an electron to be ejected at an angle θ to the direction of propagation of the incident radiation is proportional to $\sin^2 \theta$; thus the most probable direction is at right angles to the path of the incident radiation. For a given geometry, this will result in an energy spread between some minimum and maximum energy. Therefore a curve of ion current versus voltage should indicate a plateau at voltages lower than that necessary to retard the electrons of minimum energy. In this region, all the ions and a fraction of the electrons are collected. At a voltage high enough to retard the electrons of minimum energy, the ion current should start to increase and continue increasing until the voltage is sufficiently high to retard all electrons at which point the true plateau should start. The bottom two curves in Fig. 8.8 show typical curves for krypton and neon illustrating this point. The arrows

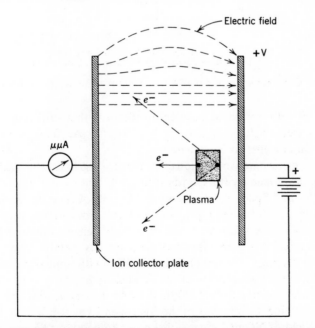

Fig. 8.9 Typical electron trajectories for which the electrons will strike the ion collector plate if a sufficiently high retarding potential is not used (Samson [2]).

indicate the values of the calculated ion chamber voltages necessary to provide the minimum and maximum voltages to retard the electrons. This problem could be essentially eliminated by using a cylindrical repeller plate with a thin central wire to collect the ions.

It is necessary to mount the collector electrode to a grounded mount via a good insulator that has good vacuum properties as well (namely, teflon, glass, and ceramics). The electrical ground acts as a shield from the leakage current which is present across the repeller plate insulator. For

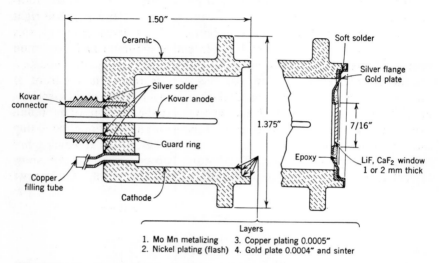

Layers
1. Mo Mn metalizing   3. Copper plating 0.0005″
2. Nickel plating (flash)   4. Gold plate 0.0004″ and sinter

Fig. 8.10 Cross section of the NASA ion chamber (courtesy A. Stober et al. [9]).

example, 20 volts across an excellent insulator with a surface resistance of 10^{12} ohms yields a leakage current of 2×10^{-11} A. This is comparable with the ion currents to be measured.

Compact nitric oxide ionization chambers for use in rockets and satellites have been developed by the Goddard Space Flight Center of the National Aeronautics and Space Administration. Details of their design have been described by A. Stober et al. [9]. Figure 8.10 shows a cross section of the ion chamber. It is fabricated out of a high density alumina shell, the cylindrical interior of which is coated with various metallic layers finishing with a gold plating. The ion chamber is filled with approximately 20 torr of NO. Thus total absorption of the incident radiation occurs. These ion chambers are sensitive from 1340 to 1040 Å; however, they are normally used to determine the intensity of the hydrogen Lyman-α line at 1215.7 Å. Since the yield of NO at Lyman-α is known to be about 85 per cent [1], it would be sufficient to measure the transmission of the lithium fluoride window at Lyman-α to provide a calibrated detector. This process is unreliable, however, since the transmission of the window is subject to change, especially in the presence of water vapor. Furthermore, permanent changes in the gas may take place. It has been found necessary, therefore, to calibrate a completed unit against some standard. Because the shelf life of a sealed off unit is uncertain, it is always necessary to recalibrate them before use. MgF_2 is less affected by water vapor than LiF, and with a transmission cut-off at 1120 Å, it is a superior window for a Lyman-α detector. A MgF_2 window 2 mm thick can transmit approximately 50 per cent of incident Lyman-α radiation. Thus the maximum efficiency of the ion chamber will be about 42 per cent. The compactness of these ion chambers makes them very convenient detectors for both laboratory and space experiments.

Photon Counters

Lukirskii, Rumsh, and Smirnov [3] have pioneered the use of Geiger counters for measuring the absolute intensity of radiation between 15 and 120 Å. Ederer and Tomboulian [4,10] have extended this work to show that the Geiger counter can also be used between 100 and 300 Å to measure absolute intensities. Caruso and Neupert [11] have used a commercial proportional counter to determine the absolute intensity of C_K radiation at 44 Å.

To operate as an absolute detector, the Geiger counter or proportional counter must produce a count for each photon transmitted by the window. In a normal Geiger counter, the production of an electron within the sensitive volume generally produces a count. Thus, if the gas pressure in the counter is high enough to totally absorb all the incident photons, and if each absorbed photon produces at least one electron, then from the transmittance of the window, the absolute photon flux can be measured. In the wavelength region shorter than 300 Å, the absorption process does lead to ionization. However, if a counter is not constructed properly, it is possible to create a "dead space" between the window and the sensitive volume such that an electron will not produce a pulse. The window is, therefore, situated nearly flush with the cylindrical sensitive volume. Figure 8.11 shows the construction used by Ederer and Tomboulian. The window was made from two layers of zapon, each 500 Å thick, that were supported and glued onto a fine mesh. The average transmittance between 100 and 300 Å of a film of zapon 1000 Å thick is about 60 per cent (see Chapter 6, Fig. 6.8). Thus, for a 60 per cent transparent screen, the efficiency of the counter will be about 36 per cent. Lukirskii et al. used a double layer of celluloid giving a total thickness of 1000 Å. The transmittance of their window is tabulated in Table 8.2. If no "dead space"

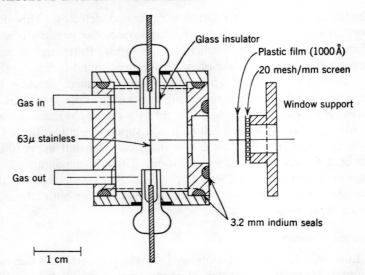

Fig. 8.11 Diagram of a photon counter suitable for absolute intensity measurements (courtesy D. L. Ederer and D. H. Tomboulian [4]).

exists, then the number of counts per sec should increase with the pressure of the gas in the counter until all the incident photons are absorbed. A further increase in the gas pressure should not change the count rate.

Various gas mixtures have been used for the counter gas, such as 80 per cent argon with 20 per cent methane, 90 per cent argon with 13 per cent absolute ethyl alcohol, and 96 per cent helium with 4 per cent isobutane. Typical total pressures are about 100 torr. The Geiger counter should always be operated on the plateau region of the counting rate *vs.* voltage curve. A typical curve is shown in Fig. 8.12 for a helium-isobutane mixture at a pressure of 70 torr. Although these gas mixtures allow the counter to be operated in the self-quenched mode, better plateau character-istics are obtained and spurious counts are decreased using electronic quenching. The dead time of an electronically quenched counter is about

Table 8.2 Transmittance of a Celluloid Film 1000 Å Thick[a]

Source Line	Wavelength (Å)	Transmittance (%)
O_K	23.3	88
C_K	44.0	92
B_K	67.0	81
Be_K	113.0	55

[a] Lukirskii et al. [3].

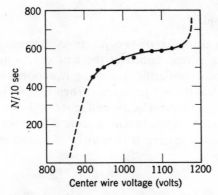

Fig. 8.12 A typical counting rate *vs.* anode voltage plateau curve. The counter was electronically quenched and the pressure of the helium isobutane gas mixture was 70 torr (courtesy D. L. Ederer and D. H. Tomboulian [4]).

200 μsec. To insure that the plateau characteristics stay constant, it is necessary to flow the filling gas continually.

Corrections for missed counts caused by the dead time τ of the counter can be made using the equation

$$N_0 = \frac{N}{1 - \tau N},$$ (8.9)

where N_0 is the true counting rate, and N is the observed counting rate.

Lukirskii et al. and Ederer and Tomboulian have investigated the possible effects of photoelectrons ejected from the window into the sensitive region of the counter. Their conclusions are that spurious counts from this source are negligible and that Geiger counters can be used as absolute detectors.

8.3 THERMAL DETECTORS

A heat sensitive detector is probably the only type of detector which has a constant response per unit incident energy from the infrared to the extreme ultraviolet. The receiving element is blackened to absorb all of the incident radiation. The blackening technique is important if a constant efficiency is to be obtained over a wide wavelength range. The degree of blackness of a receiver to extreme ultraviolet radiation is discussed at the end of this section.

The most important thermal detectors are the following:

1. The thermocouple, which produces a thermoelectric voltage;

2. The bolometer, which is characterized by a change in the resistance of the receiving element.

3. The Golay cell, which is essentially a gas thermometer.

Radiation Thermocouples

A thermocouple is made up of two junctions between two dissimilar metals. One junction is maintained at the ambient temperature and is called the cold junction, while the other one is exposed to the incident radiant flux and is called the hot junction. When the two junctions are at different temperatures, a thermoelectric emf is set up between them. This emf is proportional to the change in temperature ΔT between the junctions when ΔT is small, which in turn is proportional to the incident energy.

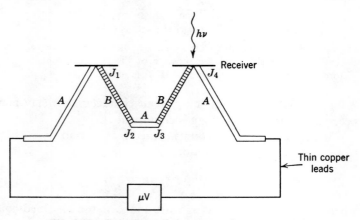

Fig. 8.13 Compensated thermocouple.

Figure 8.13 shows two couples mounted in opposition. When carefully constructed, the thermoelectric emf of each couple will be the same when no radiation is incident on the active receiver. This type of construction is known as a compensated thermocouple and is essential in dc operated circuits to prevent undue drift. The elements A and B represent the two dissimilar metals. A single thermocouple is then represented by ABA with J_1 (or J_4), a hot junction, and J_2 (or J_3), a cold junction.

If in Fig. 8.13 the two couples had been connected in reverse such that alternate junctions were hot and cold, the resulting arrangement would be called a thermopile. One might expect greater sensitivity from such a grouping of couples; however, fundamentally there is no gain in using a large number of couples [12].

Perhaps the most sensitive thermocouples are those using semiconducting materials since they have very high thermoelectric powers and low thermal conductivity. This construction is known as a Schwartz thermopile* after E. Schwartz, who made the first successful thermopile out of

* Commercially available from Hilger and Watts Ltd., England.

semiconducting compounds. Details concerning methods of manufacture and performance characteristics have been given by Brown et al. [13]. Figure 8.14 shows a Schwartz thermopile using two thermoelectric materials, one having a positive and the other a negative thermoelectric power with respect to gold. The semiconductors are mounted on gold pins, the connections forming the cold junctions. The hot junctions are made

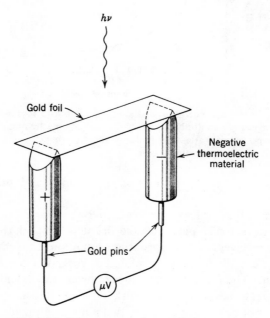

Fig. 8.14   Schwartz-type thermopile.

with the blackened gold foil receiver. Since the thermoelectric emf's of the two materials are of opposite signs with respect to gold, they are additive.

This type of construction is somewhat more robust than the bimetallic couples. Ultimate sensitivity of the order of 5×10^{-11} watts have been achieved. This sensitivity is achieved when the thermopile is under vacuum and corresponds to about 10^7 photons per second at a wavelength of 400 Å. The effect of the thermal conductivity of air is to reduce the sensitivity of a thermocouple by about a factor of 10. The speed of response of a couple, however, is reduced in a vacuum.

Figure 8.15 presents a curve of the power and energy available in a single photon as a function of wavelength. Typical sensitivities of a thermocouple range from 1 to 70 μv per μw, depending on the area of the receiving

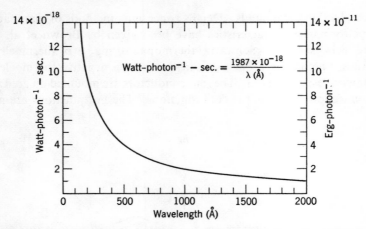

Fig. 8.15 Power and energy available in a photon as a function of wavelength. For a given wavelength, multiply the left-hand ordinate by the number of photons per sec. to obtain power in watts, and multiply the right-hand ordinate by the number of photons to obtain the energy in ergs.

elements, the total resistance of the couple, and other constructional details.

The metallic "blacks" with which the receivers of a thermopile are coated are actually pure metallic droplets which have been evaporated onto the receivers under appropriate conditions to form spherical droplets of suitable radii to provide minimum reflection of visible and infrared radiation. The receivers thus appear black and essentially absorb all of the incident radiation. The work of Harris et al. [14] on the optical properties of gold "black" shows that a highly absorbing deposit of gold is obtained for the visible and infrared when the gold spheroids have a diameter of the order of 100 Å. For vacuum uv radiation, the wavelength are reduced by about an order of magnitude or more; thus the spherical drops will appear larger to this type of radiation, and, presumably, the blackness does not necessarily hold in this region of the spectrum. In fact, Harris does find that the reflected light at 4000 Å is slightly greater than in the red. Johnston and Madden [15] have made a detailed study of the "blackness" of thermopiles. For a particular sample of gold black, they found that the integrated scattered intensity was 2.1 per cent at 584 Å, 2.0 per cent at 1025 Å, 1.9 per cent at 1216 Å, and 1.5 per cent at 4000 Å. Similar results were obtained with other samples, showing an increase in the total scattered radiation as the wavelength decreased. It is not known whether the scattering continues to increase at wavelengths shorter than 584 Å, or if the decrease in the reflectance of gold compensates for any increase in the scattered intensity. However, since the amount of scattered radiation lost is small and is very similar from the visible to 584 Å, its contribution can be neglected whether the thermopile is calibrated by visible radiation or by vacuum uv radiation.

An effect peculiar to the vacuum uv region and for radiation shorter than about a 1000 Å in particular is the photoelectric effect. As can be seen from Chapter 7, the photoelectric yield of most metals lies between

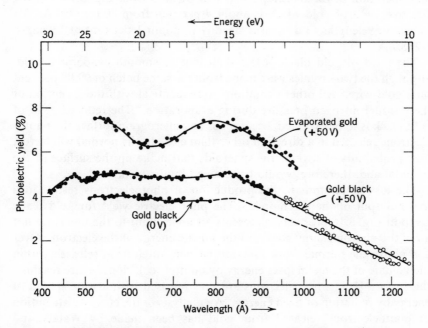

Fig. 8.16 Photoelectric yield of gold black and of a highly reflecting coating of gold. The yields are shown for zero electron collector voltage (0 V) and for a collector voltage of plus fifty volts (+50 V) (Samson [2]).

5 and 20 per cent at wavelengths between 250 and 1000 Å. This certainly means a loss of photons for the production of heat and thus an error in the determination of the absolute energies. To determine the possible effects of the photoelectric process on the performance of a thermopile, the photoelectric yield of gold black was measured by Samson between 400 and 1216 Å [2]. The gold black was prepared by evaporating pure gold onto a microscope slide in an atmosphere of nitrogen at a pressure of approximately 1 torr. The resulting coating was extremely black to the eye with a velvet-like texture. Figure 8.16 shows the yield of gold black with and without any collector voltage and, for comparison, the yield of pure gold evaporated at a pressure of 10^{-5} torr. The evaporated gold had

a mirror-like surface. The open circle data points were taken relative to a photomultiplier, sensitized with a fresh coating of sodium salicylate, then normalized to fit the curve. The remaining data points were all taken relative to the rare gas ionization chambers. The important yield curve as far as the effects on the thermopile are concerned is that of gold black with zero collector voltage since this is the condition under normal operation of the thermopile. It can be seen from Fig. 8.16 that the effective yield of gold black is about 4 per cent from 800 to 450 Å. To longer wavelengths, it decreases steadily to about 1 per cent at Lyman-α (1215.7 Å).

The yield of gold black is less than that of smooth evaporated gold, although the two samples were made from the same batch of 99.99 per cent pure gold wire. All other conditions were identical with the exception of the residual nitrogen pressure during evaporation. The reduced yield of gold black is to be expected, because for any micrograin surface the photoelectrons released in a direction other than close to the normal will have a high probability of striking the spheroids that make up the surface of the material and, therefore, would be retained by the metal.

The loss of photons in the production of photoelectrons results in a lower output of the thermopile. The photoelectric yield for gold black given in Fig. 8.16 probably represents an upper limit to the loss since not all of the photon energy goes into the kinetic energy of the electron. Over and above the normal work function barrier, an electron released within the volume of the metal loses energy by multiple collisions before reaching the surface [16]. Photoelectrons are thus released with a wide spread of energy by radiation of fixed energy. An analysis of the energy distribution of photoelectrons released from gold has been made by Walker and Weissler [17]. They find, for example, that for radiation of 704 Å (17.6 eV), nearly all of the photoelectrons have energies uniformly distributed from zero to 7.5 eV.

To verify directly the effect of photoelectrons on the thermopile output, a retarding potential of −90 volts per cm was applied between the thermopile elements and its surroundings. Using the 584 Å radiation, an increase or approximately 2 per cent in the output was detected with the retarding field on. No effect was observed at 1216 Å. Johnson and Madden [15] found for their particular thermopile an increase of 4 per cent at 584 Å and about 2 per cent at 1216 Å with either a retarding potential or a magnetic field of 1600 G to return the photoelectrons back into the thermopile elements.

Thus a thermocouple has a constant energy response from the visible to at least 584 Å, provided a retarding potential is used to return photoelectrons when appropriate.

Bolometers

The change in the electrical resistance of a material depends on its temperature coefficient of resistance α. For most metals, α is positive and equals $1/T°K$ over a wide temperature range. Therefore, $\alpha \sim 0.0035$ per °K at room temperature. The change in resistance ΔR for a small temperature change ΔT is given by

$$\Delta R = \alpha R \, \Delta T. \qquad (8.10)$$

This relation also applies to semiconductors; however, α is negative and equals $-C/T^2$, where C is a constant. Values of α for semiconducting materials are much higher than for metals, typically about 0.03 per °K at room temperature. That is, the resistance decreases by about 3 per cent per degree change in temperature.

If the resistors in two arms of a Wheatstone bridge are made from thin metallic or semiconducting strips, the bridge can be balanced when no excess of radiation falls on any one strip. On exposing one element to the incident radiation, the change in temperature and hence in resistance of the element will unbalance the bridge. A measure of this unbalanced current is a measure of the intensity of the incident radiation.

Semiconducting bolometers available commercially are known as thermistors. From (8.10) it would appear that thermistors with their higher values of α would be much more sensitive than metallic bolometers. However, the ultimate limit of detection for an ideal semiconducting bolometer is about the same as for a metal bolometer. Furthermore, the ultimate sensitivity of a bolometer is not, at present, greater than that of a thermocouple. Other features, however, may make the thermistor more attractive than thermocouples, such as their high speed of response, high resistance, and ruggedness.

Probably the most sensitive thermal detector is the superconducting bolometer. At the transition temperature, the resistance of a super-conductor suddenly falls to near zero. This means that in the vicinity of the transition point, the temperature coefficient of resistance is extremely high. For niobium nitride, $\alpha \sim 20$ per °K at a temperature of 15.5°K. Minimum detectable energies of 3×10^{-12} watts have been reported by Hulbert and Jones [18]. The need for liquid hydrogen or helium cooling tends to make their use cumbersome.

The Golay Cell

This type of heat detector is basically a differential gas thermometer. The incident radiation is absorbed by a thin membrane contained in a small gas filled cell. The heating of this receiving membrane causes expansion of the gas in contact with it, which in turn causes a deflection of a detecting membrane. This deflection, which is a measure of the incident radiant energy, is detected by its defocusing effect on a light beam falling onto a photocell. The basic construction of the Golay cell is shown in Fig. 8.17. The device was perfected by M. Golay in 1947 [19].

The Golay cell has a fast response time. Values as rapid as 600 μsec can be achieved when the cell is filled with helium. Under conditions for maximum sensitivity, the Golay cell approaches the ultimate sensitivity of the thermopile. The spectral range for this type of instrument is necessarily limited by its need for a window. With a lithium fluoride window, 1040 Å represents the short wavelength limit of sensitivity.

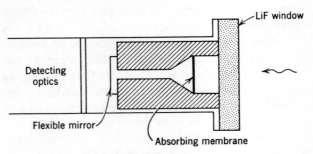

Fig. 8.17   The Golay cell.

For operation at frequencies of the order of a few cycles per sec, the thermocouple is the best uncooled thermal detector available at present. The minimum detectable radiant flux is approximately 5×10^{-11} watt, which from Fig. 8.15 corresponds to 10^7 photons per sec at 400 Å. Ionization chambers, however, can detect about 10^5 photons per sec, and with gas multiplication, about 10^3 photons per sec.

The calibration of thermal detectors is based on the known radiant flux emitted from a National Bureau of Standards lamp. The lamp consists of a tungsten filament which approaches a black body radiator. The radiant flux density emitted by these lamps at a distance of two meters is about 5×10^{-5} watt per cm². When the detector cannot be used with a window (e.g., below 1100 Å), its sensitivity in a vacuum must be measured. This consists of measuring the sensitivities of the detector in vacuum and in air with any type of constant source for the measurement. Such vacuum to air ratios can be as much as 20 to 1.

Although the use of a Standard lamp has been the conventional method in the past for calibrating thermopiles, bolometers, etc., the use of a rare gas ionization chamber provides a more direct standard. In many instances, the thermal detector can be calibrated *in situ*.

8.4 SYNCHROTRON RADIATION

The discussion of synchrotron radiation in Chapter 5 pointed out that the absolute intensity of the radiation emitted by orbiting electrons in a synchrotron could be precisely calculated from the classical theory developed by Schwinger [20]. Schwinger has shown that the instantaneous power radiated by a single electron into all wavelengths and over all angles is given by

$$P(t) = \tfrac{2}{3}\beta^3 \left(\frac{\omega_0 e^2}{R}\right)\left[\frac{E(t)}{mc^2}\right]^4,$$ (8.11)

where ω_0 and $E(t)$ are, respectively, the instantaneous angular velocity and energy of the electron orbiting in a circle of radius R. β is the ratio of the electron velocity v to the velocity of light c. In (8.11), $P(t)$ is measured in ergs per sec, e in electrostatic units, and R in cm. For high energy electrons ($\beta \approx 1$), (8.11) becomes

$$P(t) = 6.77 \times 10^{-3}\frac{(E_{BeV})^4}{R_{cm}^2} \qquad \text{watts per electron.}$$ (8.12)

where the instantaneous energy of the electron is measured in units of 1 BeV $= 10^9$ eV, and the orbital radius is measured in centimeters.

The average power radiated into all wavelengths during the acceleration interval T is given by

$$\bar{P} = \frac{1}{T}\int_0^T P(t)\, dt.$$ (8.13)

If we assume that the time taken by the electron to gain an energy in excess of 1 MeV is short compared to T, then β^3 can be taken as unity over the major portion of the integration in (8.13). Then for the case when $E(t) = E_m \sin^2(\pi t/2T)$, we find

$$\bar{P} = 0.18\left(\frac{\omega_0 e^2}{R}\right)\left(\frac{E_m}{mc^2}\right)^4,$$ (8.14)

where E_m is the maximum energy obtained by the electron during the quarter cycle acceleration interval, and where $\omega_0 = c/R$. For the Cambridge electron accelerator $E_m = 6$ BeV and $R = 2.626 \times 10^3$ cm Substituting these values into (8.14), and assuming that there are 2×10^{11} electrons in the beam, the average power radiated during the acceleration interval over all wavelengths is 69.4 kW. Since the radiation is emitted into the horizontal plane with a very small angular height, a detector of width w located at a distance d along a tangent to the orbit will be irradiated with a flux given by $(w/2\pi d)\bar{P}$, assuming the height of the detector is sufficient to intercept all the radiation. If $w = 1$ mm and $d = 10$ m, the average power received by the detector in the above example is about 1 watt. These sample calculations apply only to windowless detectors since they include radiation from the far infrared to the hard x-ray region. The use of a window would, of course, reduce the total detectable energy by many orders of magnitude. In practice, a filter of known transmittance would probably be used. The photon and energy flux per unit angstrom band are shown in Chapter 5, Figs. 5.4a,b and 5.5. It can be seen that a single monoenergetic electron of 1 BeV energy radiates about 3×10^3 photons per second per unit angstrom band at 100 Å. Thus for a beam of electrons containing 2×10^{11} electrons, the flux is 6×10^{14} photons per second for a 1 Å bandpass.

The practical use of synchrotron radiation has been described by several authors [21–24]. From the experimental difficulties described by them, it is apparent that the use of synchrotron radiation as a standard source is not straightforward. After the experimental difficulties have been surmounted, the limiting accuracy in absolute intensity measurements will depend mainly upon the accuracy in determining the number of radiating electrons, which at present is about 5 per cent, and in determining the electron energy. Although the energy can apparently be measured to a precision of 0.5 per cent, the error in the power radiated will be about 3.5 per cent since the power radiated per angstrom varies approximately as the seventh power of the electron energy.

8.5 LINE RATIOS

The intensity of a spectral line is defined as the energy radiated per second by a source. For a transition from the state n to the state m, the emission intensity I_{nm} depends on the transition probability for spontaneous emission A_{nm}, and on the number of atoms N_n in the initial state. That is,

$$I_{nm} = N_n A_{nm} h\nu_{nm}, \qquad (8.15)$$

where $h\nu_{nm}$ is the energy of each quantum of frequency ν_{nm} emitted in the transition.

If we consider two spectral lines originating from a common upper level n and making transitions to lower states m and l, as shown in Fig. 8.18, then the intensity ratio of the two lines will be

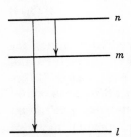

Fig. 8.18 Two transitions originating from a common upper level n.

$$\frac{I_{nm}}{I_{nl}} = \frac{A_{nm}\nu_{nm}}{A_{nl}\nu_{nl}}. \qquad (8.16)$$

Thus if the transition probabilities are known as well as the intensity of one of the lines, then the intensity of the unknown can be determined. This is known as the branching ratio method and is valid for optically thin sources. For use in the vacuum uv spectral region, a typical pair of lines would be the Lyman-β line at 1026 Å and the Balmer-Hα line at 6563 Å. Both are atomic hydrogen lines originating from the $n = 3$ level. To obtain the true ratio of the emission lines, great care must be taken in the method of exciting the light source. For example, the Lyman-β line is a resonance line and as such will be reabsorbed readily by neutral atomic hydrogen in the light source. To avoid this loss in the intensity of the 1026 Å line, the source must be operated with as little hydrogen as possible. A complication arises if the upper level exhibits fine structure. If this is the case, then the transition probabilities used in (8.16) must be averaged over the various sublevels and weighted by the statistical weight of each contributing level. The statistical weight g_n of a term n with total angular momentum J is equal to $2J + 1$, that is, to the number of levels into which the term would be split under the action of a magnetic field. Once again, however, the method of exciting a line pair is important in producing the theoretical ratio. For example, Herzberg [25] has observed that lines of hydrogen-like spectra originating from S-states are often stronger than expected on the basis of statistical population of the fine structure. This is apparently due to the long lifetime of the S-states. For the case of the $4\,^2S$-state of He II, Herzberg observed that the overpopulation disappeared if the pressure of the discharge lamp was raised to a few torr. This procedure would contradict the requirement for a low pressure discharge when a resonance line was involved. Since the nonstatistical distribution can be removed by increasing the collision frequency of electrons with the long lived states, it would appear as if discharges with axial magnetic fields such as a duoplasmatron would be the most suitable in producing statistical distribution among the fine structure levels.

The use of line ratios to measure absolute intensities in the vacuum uv and to calibrate monochromators was first described by Zaidel et al. [26] and later by Gladushchak and Shreider [27]. They selected line pairs from Al III and Si IV to avoid the problem of self-absorption. The fine structure levels were resolved in this case. Moreover the transition probabilities were calculated using the tables of Bates and Damgaard [28]. Independently and about the same time as Zaidel et al., Griffin and McWhirter [29] described the same technique for intensity measurements. They made use of the 1026 and 6563 Å lines of atomic hydrogen originating from discharges in Zeta. Van Eck et al. [30] used the transitions $3\,^1P-2\,^1S$ and $3\,^1P-1\,^1S$ in neutral helium producing the 5016 and 537 Å lines. The lines were produced by firing 30 keV He$^+$ ions into a chamber filled with hydrogen or neon. Many of the ions were neutralized into the $3\,^1P$ state by electron capture processes during collisions with the target gas. This method of producing the He lines prevented self-absorption. The $3\,^1P$ level of He does not have the problem of fine structure. Hinnov and Hofmann [31] carried out calibrations using He II and H I line pairs. They simultaneously calibrated the monochromator and detector system.

Let us define the sensitivity $S(\lambda)$ of the monochromator and detector system for a given wavelength λ as the ratio of the detector response $i(\lambda)$ to the absolute intensity of the incident radiation which enters the monochromator. If the absolute intensity incident per unit area on a slit of width w and length l is $I(\lambda)$, then

$$S(\lambda) = \frac{i(\lambda)}{lwI(\lambda)}.\tag{8.17}$$

For two different wavelengths, we have

$$S(\lambda_1) = S(\lambda_2)\frac{i_1}{i_2}\frac{I_2}{I_1}.\tag{8.18}$$

Using a suitable light source, the ratio I_2/I_1 for two lines originating from the same upper level can be calculated from (8.16); the ratio i_1/i_2 is observed from the response of the detector to the two lines; and finally $S(\lambda_2)$ is determined by calibrating the monochromator-detector system using the known spectral emission from a standard tungsten ribbon lamp whose short wavelength limit is determined by its glass envelope. Thus the sensitivity $S(\lambda_1)$ of the monochromator-detector system is known for a given wavelength in the vacuum uv. This technique is convenient for calibrating monochromator systems; however, the calibration is limited to only a few selected wavelengths.

8.6   LYMAN-ALPHA INTENSITY MEASUREMENTS

Single photon coincidence method.   Cristofori et al. [32] have described an ingenious method for calibrating the response of a detector system to radiation of 1216 Å. Consider collisions between a beam of electrons of 2 keV energy and molecular hydrogen. Many of the collisions will not only dissociate the molecules into atomic hydrogen, but will leave the atoms in various excited states. Some of the atoms excited into the $n = 3$ level will decay with two successive transitions: to the $n = 2$ level, emitting a photon of 6563 Å; and then to the $n = 1$ level, emitting a photon of 1216 Å. If the two photons are detected in coincidence, the ratio between the number of coincidences and the number of detected photons of 6563 Å immediately gives the value of the efficiency of a 1216 Å detector, which in turn, of course, gives the absolute flux of the 1216 Å radiation.

To clarify the steps involved in such a calibration, consider the following analysis. Let N_3 be the number of atoms that, when excited to the $n = 3$ level, decay into the $n = 2$ level. Furthermore, let N_2 be the number of atoms excited directly into the $n = 2$ level (see Fig. 8.19). Then the number of counts produced by the 6563 and 1216 Å photons is given by:

$$n_{6563} = N_3\xi_r\Omega_r,\tag{8.19}$$

and

$$n_{1216} = (N_3 + N_2)\xi_u\Omega_u,\tag{8.20}$$

where ξ is the efficiency of the detectors, and Ω is the fraction of the total number of photons emitted which intercept the detectors; that is, it is the

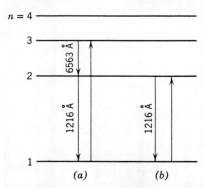

Fig. 8.19 Transitions in atomic hydrogen illustrating the spontaneous emission of 1216 Å and the emission of 1216 Å in coincidence with 6563 Å.

fractional solid angle observed by the detectors. The number of coincidences n_c will be

$$n_c = N_3\xi_r\xi_u\Omega_r\Omega_u.\tag{8.21}$$

Taking the ratio of (8.21) and (8.19), we get

$$\frac{n_c}{n_{6563}} = \xi_u\Omega_u,\tag{8.22}$$

where Ω_u is known from the geometry of the system. Thus a measure of the ratio of coincidence counts to the number of counts produced by the 6563 Å photons gives the efficiency of the 1216 Å detector. Substituting this value into (8.20) and measuring the number of counts produced by the 1216 Å photons gives $(N_3 + N_2)\Omega_u$, the absolute number of 1216 Å photons radiated into the detecting system.

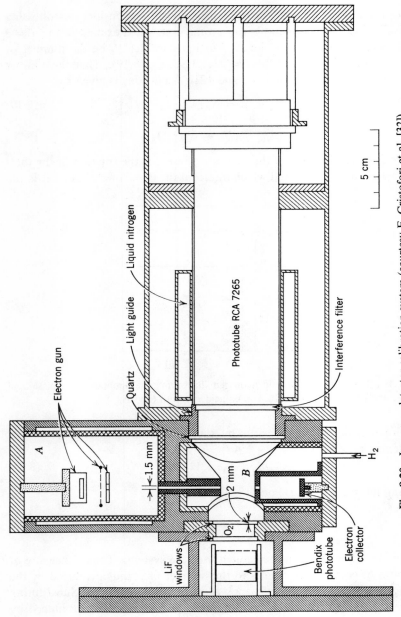

Electron gun

A

1.5 mm

2 mm

B

O₂

LiF windows

Quartz — Light guide

Liquid nitrogen

Phototube RCA 7265

Interference filter

H₂

Bendix phototube

Electron collector

5 cm

Fig. 8.20 Lyman-α detector calibration system (courtesy F. Cristofori et al. [32]).

Two important points have not been mentioned yet. The first is that some of the transitions will be from the $3P$ to the $2S$ metastable state, which will eventually decay to the ground state. However, such transitions will not be counted as coincidences. The metastable states, however, can be de-excited by creating an electric field of several 100 volts per cm in the vicinity of the atoms. The second point is that of spurious coincidence counts $n_{c.sp.}$. The number of these is given by [33]

$$n_{c.sp.} = 2n_{1216}n_{6563}\tau, \tag{8.23}$$

where τ is the resolution time of the coincidence. Thus the signal to noise ratio is given by the ratio of (8.21) and (8.23), namely,

$$\frac{n_c}{n_{c.sp.}} = \frac{1}{2(N_3 + N_2)\tau}. \tag{8.24}$$

In order that this ratio may be large enough, τ must be as small as possible.

Figure 8.20 shows the experimental arrangement used by Cristofori et al. In order to isolate the 6563 Å radiation, they used a narrow band interference filter which had a maximum transmission of 36 per cent centered on 6560 Å. The filter had a band pass of about 140 Å. The 6563 Å detector was chosen as one having a large signal to noise ratio compatible with as large a diameter as possible. To further increase the signal to noise ratio, the photomultiplier cathode was cooled by liquid nitrogen.

The 1216 Å detector shown in Fig. 8.20 is the Bendix magnetic multiplier, which uses a pure tungsten cathode. With LiF windows and an oxygen filter, the spectral response of this detector is very limited.

With the high voltages used on the cathodes of the two detectors (typically 2000 V negative on the Bendix cathode), it was apparently unnecessary to apply a further field to de-excite the metastable atoms. The magnetic field of the Bendix multiplier possibly also contributed to their de-excitation.

The electron beam was accelerated to about 1800 eV before entering the experimental chamber B containing hydrogen at a pressure of 2×10^{-3} torr. The electron current collected was about 0.02 uA.

Using the above experimental technique, Cristofori et al. were able to measure the absolute efficiency of their 1216 Å detector to an accuracy of approximately 20 per cent.

Black Body Radiation. A Lyman-alpha black body radiator has been constructed by Boldt [34]. With a cascade arc burning in argon at a temperature of 11,900°K at its axis, sufficient hydrogen was added to make the Lyman-α line, viewed end-on, as intense as the black body radiation of a source at the above temperature. Self-reversal of the Lyman-α line was practically eliminated by keeping hydrogen away from the cooler region between the arc and the slit by appropriate flow of gases.

The uncertainty in the absolute intensity of the radiation is caused mainly by the uncertainty in the temperature measurement of the arc. This error is about 20 per cent.

8.7 CONCLUSIONS

An absolute detector based on the principle of photoionization is the most precise and sensitive standard available for the measurement of absolute intensities.

To measure the absolute intensity of radiation from approximately 2 to 300 Å, a properly constructed photon counter should be used. In the range 250 to 1022 Å, it is simpler and more accurate to use the rare gas ion chambers. For wavelengths longer than 1022 Å, a secondary standard with a "flat" response should be calibrated against a rare gas ion chamber. Probably the best secondary standard is the thermocouple. Calibrating the thermocouple directly with a rare gas ion chamber involves only one step compared to the three steps involved with the standard procedure. More-over, most research laboratories can easily construct standard ion chambers for the calibration of thermocouples. A freshly prepared sodium salicylate coated photomultiplier can also be used as a secondary standard from 1000 to 3500 Å with moderate accuracy. When calibrated in the vicinity of 1000 Å, the accuracy over the range 1000 to 3500 Å should be within ±20 per cent. Once the conditions for establishing a "flat" response with sodium salicylate are understood, the accuracy is likely to be greater than that of the thermocouple; however, it is probable that the salicylate coated photomultiplier would need recalibrating prior to any measurements. Thus the thermocouple probably remains the best secondary standard.

Recent reviews on absolute intensity measurements in the vacuum uv have been given by Shreider [35] and Hinteregger [36].

REFERENCES

[1] K. Watanabe, F. F. Marmo, and E. C. Y. Inn, *Phys. Rev.* **91**, 1155 (1953).
[2] J. A. R. Samson, *J. Opt. Soc. Am.* **54**, 6 (1964),
[3] A. P. Lukirskii, M. A. Rumsh, and L. A. Smirnov, *Opt. i Spektroskopia* **9**, 505 (1960); English trans. *Opt. Spectry.* **9**, 262 (1960).
[4] D. L. Ederer and D. H. Tomboulian, *Appl. Optics* **3**, 1073 (1964).
[5] F. M. Matsunaga, R. S. Jackson, and K. Watanabe, *J. Quant. Spectrosc. Radiat. Transfer* **5**, 329 (1965).

[6] R. E. Huffman, Y. Tanaka, J. C. Larrabee, *J. Chem. Phys.* **39**, 902 (1963); *Appl. Optics* **2**, 947 (1963).

[7] J. A. R. Samson, in *Advances in Atomic and Molecular Physics*, eds. D. R. Bates and I. Estermann (Academic, New York, 1966) Vol. II.

[8] A. Sommerfeld, *Wave Mechanics* (Methuen, London, 1930), p. 181.

[9] A. K. Stober, R. Scolnik, and J. P. Hennes, *Appl. Optics* **2**, 735 (1963).

[10] D. H. Tomboulian, *J. Appl. Phys.* (Japan), Suppl. 1, **4**, 542 (1965).

[11] A. J. Caruso and W. M. Neupert, *Appl. Optics* **4**, 247 (1965).

[12] R. A. Smith, F. E. Jones, and R. P. Chasmar, *The Detection and Measurement of Infra-Red Radiation* (Oxford U. P., London, 1960), p. 57.

[13] D. A. H. Brown, R. P. Chasmar, and P. B. Fellgett, *J. Sci. Instr.* **30**, 6, 195 (1953).

[14] L. Harris, R. T. McGinnies, and B. M. Seigel, *J. Opt. Soc. Am.* **38**, 582 (1948).

[15] R. G. Johnston and R. P. Madden, *Appl. Optics* **4**, 1574 (1965).

[16] H. Hinteregger, *Phys. Rev.* **96**, 538 (1954).

[17] W. C. Walker and G. L. Weissler, *Phys. Rev.* **97**, 1178 (1955).

[18] J. A. Hulbert and G. O. Jones, *Proc. Phys. Soc.* **B68**, 801 (1955).

[19] M. J. E. Golay, *Rev. Sci. Instr.* **18**, 347 (1947); **18**, 357 (1947).

[20] J. Schwinger, *Phys. Rev.* **75**, 1912 (1949).

[21] D. H. Tomboulian and P. L. Hartman, *Phys. Rev.* **102**, 1423 (1956).

[22] D. H. Tomboulian and D. E. Bedo, *J. Appl. Phys.* **29**, 804 (1958).

[23] Y. Cauchois, C. Bonelle, and G. Missoni, *Compt. Rend.* **257**, 409 and 1242 (1963).

[24] K. Codling and R. P. Madden, *J. Appl. Phys.* **36**, 380 (1965).

[25] G. Herzberg, *Z. Physik* **146**, 269 (1956).

[26] A. N. Zaidel, G. M. Malyshev, and E. Ya. Shreider, *Zh. Tekhn. Fiz.* **31**, 129 (1961); English trans. *Soviet Phys.-Tech. Phys.* **6**, 93 (1961).

[27] V. I. Gladushchak and E. Ya. Shreider, *Opt. i Spektroskopia* **14**, 815 (1963); English trans., *Opt. Spectr.* **17**, 75 (1964).

[28] D. R. Bates and A. Damgaard, *Phil. Trans. Roy. Soc.* (London) **242**, 101 (1949).

[29] W. G. Griffen and R. W. P. McWhirter, in *Conference on Optical Instruments and Techniques*, ed. K. J. Habell (Chapman and Hall, London, 1962) p. 14.

[30] J. Van Eck, F. J. Heer, and J. Kistemaker, *Phys. Rev.* **130**, 656 (1963).

[31] E. Hinnov and F. Hofmann, *J. Opt. Soc. Am.* **53**, 1259 (1963).

[32] F. Cristofori, P. Fenici, G. E. Frigerio, N. Molho, and P. G. Sona, *Physics Letters* **6**, 171 (1963).

[33] B. B. Rossi, *Ionization Chambers and Counters* (McGraw-Hill, New York, 1959), 1st ed., p. 238.

[34] G. Boldt, *Proc. Fifth Int. Conference on Ionization Phenomena in Gases* (North Holland, Munich, 1961), p. 925; *J. Quant. Spectrosc. Radiat. Transfer* **5**, 91 (1965).

[35] E. Ya. Shreider, *Zh. Tekhn. Fiz.* **34**, 2089 (1964); English trans., *Soviet Phys.-Tech. Phys.* **9**, 1609 (1965).

[36] H. E. Hinteregger, *Space Sci. Reviews* **4**, 461 (1965).

9

Polarization

Research with polarized vacuum uv radiation has been seriously hindered by the lack of suitable polarizers. Nearly all birefringent materials are opaque in this spectral region. However, with the increased activity in this field, a considerable amount of attention has been given to polarization. The first reported study of polarized vacuum uv radiation was by Cole and Oppenheimer [1] in 1962. They measured the degree of polarization of the radiation transmitted by their grazing incidence vacuum spectrograph in the range 304 to 1216 Å and found that the polarization increased from zero at 304 Å to 75 percent at 1216 Å. Since 1962 a considerable amount of work has been reported on the production of polarized radiation. With the exception of polarized sources of radiation such as obtained from the synchrotron (see Chapter 5), the techniques for production of polarized radiation have included most of the conventional methods used in the visible region, namely, birefringence, reflection, and refraction. These techniques are described below. When vibrations of polarized radiation are referred to, it is to be understood that these vibrations refer to the electric vector. The plane of polarization is the plane which contains the electric vector and which passes through the axis of the beam of light.

9.1 BIREFRINGENCE

Double refraction is invariably present in crystals that are not of the regular system, that is, uniaxial and biaxial crystals. When a beam of light is incident upon such crystals, it is split into two that, in general, travel with different velocities and in different directions and vibrate in particular planes. For uniaxial crystals, the beam of light that obeys Snell's law of refraction is called the ordinary ray or O ray while the other beam is called the extraordinary ray or E ray. The O vibrations are perpendicular to the principal plane of the O ray, while the E vibrations lie in the principal plane of the E ray. The principal plane is defined as the plane containing the optic axis and the ray. Thus the O vibrations are always perpendicular to the optic axis. The two indices of refraction n_E and n_O are associated with the E and O rays, respectively. The difference in the refractive indices $(n_E - n_O)$ is called the birefringence of the crystals. In calcite $(CaCO_3)$ this quantity is negative; thus, calcite is called a negative crystal. Quartz and magnesium fluoride are both uniaxial positive crystals.

Calcite has apparently been used as a polarizer down to 1900 Å as early as 1938 [2], but its properties remained relatively unknown. Müller [3]

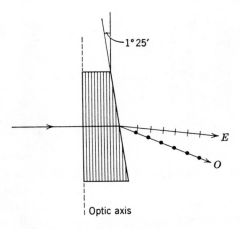

Fig. 9.1 Calcite crystal polarizer.

has recently republished information on this polarizer. Although calcite does not normally transmit below 2000 Å, it can, under certain conditions, transmit the E ray down to at least 1900 Å. In the crystal, the CO_3 ions lie in planes parallel to each other and perpendicular to the optic axis. Incident radiation polarized with the electric vector parallel to the plane containing the CO_3 ions excites the ions and is thereby absorbed, while radiation polarized perpendicular to the ion planes is not absorbed by this mechanism and is transmitted to about 1900 Å [4,5]. If a weakly wedge-shaped plate is cut from a calcite crystal with one face oriented parallel to the optic axis as shown in Fig. 9.1, then the incident radiation will emerge from the crystal split into an O and E ray with vibrations perpendicular and parallel, respectively, to the plane of the paper. It is conventional to represent the vibrations of the electric vector perpendicular to the plane of the paper by dots, and vibrations in the plane, by short lines. Since the O ray vibrates perpendicular to the optic axis, it thus vibrates parallel to the plane of the CO_3 ions and will be absorbed more rapidly than the E ray as the transmission limit of the crystal is reached. The principal indices of refraction for calcite are [6]:

λ (Å)	n_O	n_E
2000	1.903	1.577
2573	1.760	1.530
5893	1.658	1.486

Müller used a calcite plate 1 mm thick with a wedge angle of 1°25′. However, it appears unnecessary to use a wedge for radiation below 2000 Å since its only purpose is to separate the O and E rays as they

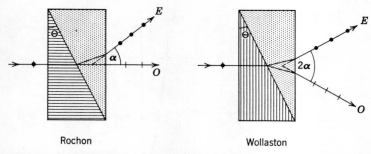

Rochon   Wollaston

Fig. 9.2 Orientation of the optic axis in the Rochon and Wollaston-type polarizer.

emerge from the crystal. Below 2000 Å, the anisotropic nature of the crystal should produce plane polarized radiation vibrating parallel to the optic axis using simply a thin parallel plate of calcite.

The successful production of a MgF_2 polarizing prism by Johnson [7] tends to make the above calcite polarizer obsolete for use in the vacuum uv. Magnesium fluoride is transparent down to about 1125 Å; however, its polarizing properties are excellent only down to 1400 or 1300 Å. Johnson reports that it is useable down to 1200 Å. The crystals can be constructed into Rochon or Wollaston-type polarizers as shown in Figure 9.2. The Rochon polarizer has the advantage that one ray (the O ray) is undeviated; the Wollaston type has the advantage that the two linearly polarized beams are separated by twice as much as the Rochon type for the same thickness of material. The MgF_2 polarizer, as constructed by Johnson and the Karl Lambrecht Company, Chicago, Ill., is of the Wollaston type and has a cross section of 6 × 20 mm, is 3 mm thick, and has a prism angle Θ of 26.6°. The two matched prisms were spaced leaving a 0.1 mm air gap, which, of course, will be evacuated with the monochromator. The gap was left because of the practical difficulties involved in making optical contact between materials as soft as magnesium fluoride. The E and O rays diverged at an angle 2α equal to 1.12° in the wavelength range 2100 to 1400 Å. Below 1400 Å, the divergence decreased going to zero at about 1200 Å. The angular separation of the beams is quoted by Johnson to be given by $\sin \alpha = (n_E - n_O) \tan \Theta$ and is thus proportional to the birefringence $(n_E - n_O)$. At 5893 Å, the birefringence of MgF_2 is 0.012 [8], and between 2100 and 1400 Å, it is 0.019 [7]. Figure

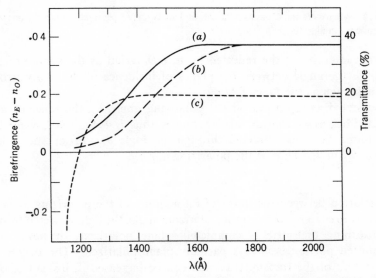

Fig. 9.3 Optical properties of magnesium fluoride: (*a*) the vertically polarized component; (*b*) the horizontally polarized component. Maximum transmittance for each component is 50 per cent. Accuracy is ±3 per cent transmittance units. Curve (*c*) is the birefringence (courtesy W. C. Johnson [7]).

9.3 shows the birefringence of the polarizer along with the percent transmittance for radiation polarized (a) vertically, that is, parallel to the entrance slit, and (b) horizontally.

9.2 POLARIZATION BY REFLECTION AND REFRACTION

When unpolarized radiation of any wavelength is reflected or refracted by a dielectric or is reflected from the surface of a metal, it is, in general, partially plane polarized. The polarization of light by reflection from a glass plate was first discovered by Malus in 1808. He showed that if light were incident at some critical angle $\bar{\theta}_i$, then the reflected light was completely plane polarized. Figure 9.4 shows the arrangement for two such mirrors. The mirror *B* acts as polarizer while mirror *C* acts as the analyzer. Malus showed that if the mirror *C* was rotated about the line *BC* as an

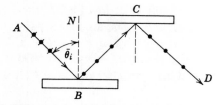

Fig. 9.4 Mirrors *B* and *C* act as polarizer and analyzer, respectively, when the angle of incidence is equal to the polarizing angle.

axis, the intensity of the reflected beam *CD* varied as the square of the cosine of the angle between the planes of incidence of the two mirrors. This is known as the *Law of Malus*.

Brewster discovered that at the polarizing angle, $\bar{\theta}_i$, the reflected and refracted beams were always at right angles, thus $\theta_i + \theta_t = 90°$, where the transmitted beam is refracted through an angle $\bar{\theta}_t$. From Snell's law $n = \sin\theta_i/\sin\theta_t$. Thus at the polarizing angle,

$$n = \tan\bar{\theta}_i. \tag{9.1}$$

This relation between the index of refraction and the polarizing angle is known as *Brewster's law*. At the polarizing angle, the light reflected from a nonabsorbing dielectric is completely plane polarized; however, the transmitted portion is always partially plane polarized. The degree of polarization of the transmitted beam can be increased if, instead of one plate, a pile of plates are used as shown in Fig. 9.5. The repeated reflections decrease the amount of perpendicular vibrations in the transmitted beam. The pile of plates shown in Fig. 9.5 are stacked in two opposing groups such that the incident and emergent beams lie on the same axis.

The degree of polarization *P* is defined as

$$P = \left| \frac{I_\perp - I_\parallel}{I_\perp + I_\parallel} \right|, \tag{9.2}$$

where $I_\parallel$ and $I_\perp$ are the intensities of the components vibrating parallel and perpendicular to the plane of incidence, respectively, and can refer to

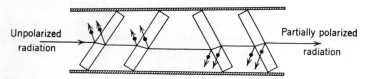

Unpolarized radiation

Partially polarized radiation

Fig. 9.5 A pile of plates polarizer, constructed to preserve the direction of the incident beam.

either the reflected or transmitted beams. The degree of polarization can be calculated using Fresnel's equations for a dielectric, namely,

$$R_\perp = \frac{\sin^2 (\theta_i - \theta_t)}{\sin^2 (\theta_t + \theta_i)}, \tag{9.3a}$$

$$R_\| = \frac{\tan^2 (\theta_i - \theta_t)}{\tan^2 (\theta_i + \theta_t)}, \tag{9.3b}$$

$$T_\perp = \frac{\sin 2\theta_i \sin 2\theta_t}{\sin^2 (\theta_i + \theta_t)}, \tag{9.3c}$$

$$T_\| = \frac{\sin 2\theta_i \sin 2\theta_t}{\sin^2 (\theta_i + \theta_t) \cos^2 (\theta_i - \theta_t)}, \tag{9.3d}$$

where R is the reflectance, and T is the transmittance. The above equations refer to a single surface. If, however, a beam of light passes through the two surfaces of a plate and we neglect multiple reflections, the transmitted light is proportional to T^2, and for m plates, proportional to T^{2m}. Thus the degree of polarization after transmission of m plates by radiation incident at some arbitrary angle θ_i is given by

$$P(m, n, \theta_i) = \frac{1 - \cos^{4m} (\theta_i - \theta_t)}{1 + \cos^{4m} (\theta_i - \theta_t)}, \tag{9.4}$$

where $n = \sin \theta_i / \sin \theta_t$. At the polarizing angle (9.4) becomes

$$P = \frac{1 - a^{4m}}{1 + a^{4m}}, \tag{9.5}$$

where $a = 2n/(1 + n^2)$. When multiple reflections are included, they will of course, tend to decrease the degree of polarization. It can be shown [9] that in this case the values of $T_\perp$ and $T_\|$ to be used in (9.2) for m plates are given by

$$T_\perp = \frac{\tan^2 (\theta_i + \theta_t) - \tan^2 (\theta_i - \theta_t)}{\tan^2 (\theta_i + \theta_t) + (2m - 1) \tan^2 (\theta_i - \theta_t)}, \tag{9.6}$$

and

$$T_\| = \frac{\sin^2 (\theta_i + \theta_t) - \sin^2 (\theta_i - \theta_t)}{\sin^2 (\theta_i + \theta_t) + (2m - 1) \sin^2 (\theta_i - \theta_t)}. \tag{9.7}$$

At the polarizing angle, this gives

$$P = \frac{m}{m + [2n/(n^2 - 1)]^2}. \tag{9.8}$$

Weinberg [9] has calculated the degree of polarization for several plates and indices of refraction using the rigorous equations above. Table 9.1 shows his calculations for a material with an index of refraction $n = 1.621$ and for $m = 1, 2, 5,$ and 10. From (9.5) and (9.8) it can be seen that P increases as the index of refraction increases.

Table 9.1 Calculated Degree of Polarization of Transmitted Light using a Pile of Plates Polarizer with 1, 2, 5, and 10 Plates and Index of Refraction $n = 1.621$ for Single and Multiple Reflections[a]

θ_i (deg)	θ_t (deg)	Singly Reflecting				Multiply Reflecting			
		1	2	5	10	1	2	5	10
00	00	0	0	0	0	0	0	0	0
05	03.08	0.00112	0.00224	0.00560	0.0112	0.00106	0.00192	0.00372	0.00542
10	06.15	0.00452	0.00904	0.0226	0.0452	0.00428	0.00773	0.0150	0.0219
15	09.19	0.0103	0.0206	0.0515	0.103	0.00976	0.0176	0.0342	0.0498
20	12.18	0.0187	0.0373	0.0931	0.185	0.0177	0.0319	0.0619	0.0901
25	15.11	0.0299	0.0598	0.148	0.291	0.0283	0.0511	0.0988	0.144
30	17.97	0.0444	0.0866	0.219	0.417	0.0420	0.0756	0.146	0.211
35	20.72	0.0627	0.125	0.304	0.556	0.0592	0.106	0.203	0.292
40	23.36	0.0853	0.169	0.403	0.694	0.0803	0.143	0.271	0.385
45	25.86	0.113	0.224	0.514	0.813	0.106	0.188	0.347	0.486
50	28.20	0.147	0.288	0.630	0.902	0.137	0.238	0.429	0.585
55	30.35	0.189	0.364	0.742	0.957	0.174	0.295	0.510	0.672
58.33[b]	31.70	0.220	0.421	0.808	0.978	0.201	0.334	0.557	0.715
60	32.29	0.239	0.452	0.839	0.985	0.216	0.355	0.579	0.732
65	33.99	0.299	0.549	0.912	0.996	0.264	0.413	0.625	0.755
70	35.43	0.370	0.651	0.960	0.999	0.314	0.461	0.642	0.738
75	36.58	0.453	0.751	0.985	0.999+	0.364	0.493	0.625	0.687
80	37.41	0.546	0.841	0.996	0.999+	0.408	0.501	0.581	0.614
85	37.92	0.646	0.912	0.999	0.999+	0.438	0.485	0.519	0.531
90	38.09	1.00	1.00	1.00	1.00	0.449	0.449	0.449	0.449

[a] Courtesy J. L. Weinberg, *Appl. Optics* **3**, 1057 (1964).
[b] Polarizing angle.

Walker [10] has constructed a pile of plates polarizer after the fashion shown in Fig. 9.5, using cleaved discs of LiF about 0.5 mm thick and oriented at an angle of 60°. The refractive index [11] and the polarizing angle of LiF is shown as a function of wavelength in Fig. 9.6. At 1200 Å, the polarizing angle is 58°40′, and $n = 1.64$. The degree of polarization for four plates, measured as a function of wavelength, is shown in Fig. 9.7. The solid line represents the experimental results of Walker [10] while the dashed line represents the theoretical values obtained from (9.4), that is, neglecting multiple reflections. Since the experimental value of P is less than the theoretical value, it would appear that some dilution of the polarization occurs due to multiple reflections. The dilution, however, is not so great as predicted by (9.8).

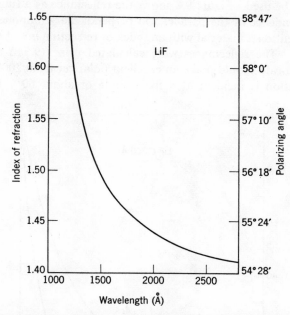

Fig. 9.6 The refractive index and polarizing angle of LiF (after Schneider [11]).

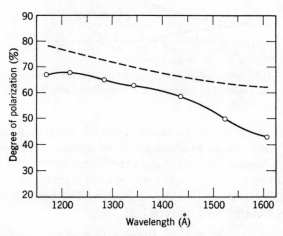

Fig. 9.7 The degree of polarization for radiation transmitted by a pile of plates polarizer with four plates. The angle of incidence was 60°. The solid line represents the experimental data by Walker [10], while the dashed line is the calculated degree of polarization neglecting multiple reflections.

303

To obtain complete polarization, the reflectance method at the polarizing angle must be used.   Figure 9.8 shows the reflectance as a function of the angle of incidence of the parallel and perpendicular components of radiation incident on a material with an index of refraction $n = 1.64$ (e.g. LiF at 1200 Å).   The reflectances were calculated using (9.3 a) and (9.3 b). Lithium fluoride should make an excellent polarizer from 2000 to 1100 Å when radiation is incident at a fixed angle of about 60°.   For shorter

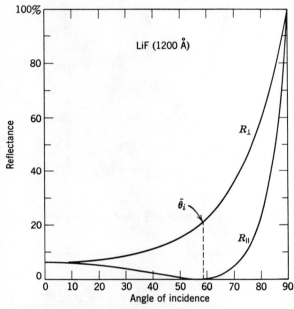

Fig. 9.8 The calculated reflectance of the parallel and perpendicular components of radiation incident on a material of refractive index equal to 1.64.   The polarizing angle is represented by $\bar{\theta}_i = 58°40'$.

wavelengths due to the lack of transparent crystals the reflectance method is the only suitable one to produce polarized radiation.   However, the radiation is only partially polarized since the reflecting material now has a complex index of refraction $(n - ik)$, where n is the real part of the index of refraction, and k is the extinction coefficient.   The generalized Fresnel equations for reflectance now become [12]

$$R_\perp = \frac{a^2 + b^2 - 2a \cos \theta_i + \cos^2 \theta_i}{a^2 + b^2 + 2a \cos \theta_i + \cos^2 \theta_i}, \tag{9.9}$$

and

$$R_\parallel = R_\perp \frac{a^2 + b^2 - 2a \sin \theta_i \tan \theta_i + \sin^2 \theta_i \tan^2 \theta_i}{a^2 + b^2 + 2a \sin \theta_i \tan \theta_i + \sin^2 \theta_i \tan^2 \theta_i}, \tag{9.10}$$

where

$$2a^2 = [(n^2 - k^2 - \sin^2 \theta_i)^2 + 4n^2k^2]^{1/2} + (n^2 - k^2 - \sin^2 \theta_i),$$

and

$$2b^2 = [(n^2 - k^2 - \sin^2 \theta_i)^2 + 4n^2k^2]^{1/2} - (n^2 - k^2 - \sin^2 \theta_i).$$

Thus, knowing the value of n and k for a particular wavelength, reflectance curves similar to those shown in Fig. 9.8 can be calculated from (9.9) and

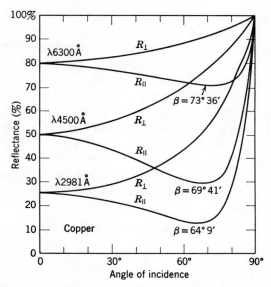

Fig. 9.9 Reflectance of copper as a function of the angle of incidence and the wavelength (courtesy F. A. Jenkins and H. E. White [13]).

(9.10). Figure 9.9 shows typical reflectance curves obtained experimentally for copper at several wavelengths [13]. The reflectance curves for a metal are very similar to those obtained from a dielectric, with the exception that the metallic reflectance is generally greater and the minimum reflectance of the $R_{\parallel}$ component does not go to zero as in the case for a dielectric.

When radiation is reflected from a metal, a change in phase occurs which is different for each of the two components $R_{\parallel}$ and $R_{\perp}$. The difference Δ between the two phase changes is more important than the values of the individual phase changes and is given by [14]

$$\tan \Delta = \frac{-2a \sin \theta_i \tan \theta_i}{a^2 + b^2 - \sin^2 \theta_i \tan^2 \theta_i}, \tag{9.11}$$

where a and b are defined in (9.9) and (9.10).

The principal angle of incidence is defined as the angle of incidence $\bar{\theta}_i$ for which the phase change is 90°. The condition for this is that $a^2 + b^2 - \sin^2 \bar{\theta}_i \tan^2 \bar{\theta}_i$ should be equal to zero, which gives

$$(\sin \bar{\theta}_i \tan \bar{\theta}_i)^4 = (n^2 + k^2)^2 - 2(n^2 - k^2) \sin^2 \bar{\theta}_i + \sin^4 \bar{\theta}_i. \quad (9.12)$$

Equation (9.12) can often be approximated by

$$\sin \bar{\theta}_i \tan \bar{\theta}_i = (n^2 + k^2)^{1/2}, \quad (9.13)$$

when $\sin^2 \bar{\theta}_i < (n^2 + k^2)$. When k goes to zero, (9.12) becomes identical with (9.1) for a dielectric. Because of the phase change Δ between the parallel and perpendicular components of the reflected light, incident plane polarized light vibrating at some angle of the plane of incidence will, in general, be elliptically polarized on reflection.

The principal angle of incidence is often assumed to be identical with (a) the angle for which $R_\parallel$ is a minimum and (b) the angle for which the degree of polarization is a maximum. However, this is not the case unless $k = 0$. The condition for $R_\parallel$ to be a minimum is obtained by differentiating (9.10) with respect to θ_i and equating the result to zero. The procedure is very tedious with various substitutions required. Humphreys-Owen [15] has solved this problem and has shown that the minimum occurs at an angle $\theta_i = \beta$ given by

$$2(p^2 + q)v^3 + p^2(p^2 - 3)v^2 - 2p^4v + p^4 = 0, \quad (9.14)$$

where $v = \sin^2 \beta$, $p = n^2 + k^2$, and $q = n^2 - k^2$.

Damany [16,17] has solved the equivalent problem for the angle $\theta_i = \varphi$ for which the degree of polarization is a maximum, obtaining

$$(1 - \alpha)t^4 + (p^2 - \alpha)t^3 + (p^2 - \alpha^2)t^2 + 4p^2\alpha t - 4p^4 = 0, \quad (9.15)$$

where $t = \tan^2 \varphi$ and $\alpha = 2q - p^2$. Table 9.2 lists the three angles $\bar{\theta}_i$, β,

Table 9.2 Values of the Principal Angle $\bar{\theta}_i$, Polarizing Angle φ, and the Angle β for which $R_\parallel$ is a minimum, for Different Materials at 584 Å. The Numerical Values of the Optical Constants were taken from References [1] and [18][a]

Angle of Incidence	Al	Ag	Au	MgF$_2$	SiO	ZnS	Al	Al$_2$O$_3$	Glass
	$n=0.77$ $k=0.09$	$n=0.92$ $k=0.33$	$n=1.07$ $k=0.71$	$n=0.93$ $k=0.33$	$n=0.80$ $k=0.47$	$n=0.79$ $k=0.20$	$n=0.71$ $k=0.018$	$n=0.74$ $k=0.56$	$n=0.83$ $k=0.45$
$\bar{\theta}_i$	38°24′	47°22′	56°04′	47°34′	49°10′	44°57′	35°25′	50°50′	48°58′
φ	38°59′	44°45′	47°02′	45°01′	44°13′	41°24′	35°20′	45°05′	44°30′
β	37°13′	42°04′	48°47′	42°23′	38°52′	38°33′	35°20′	37°57′	39°38′

[a] Courtesy H. Damany, *Optica Acta* **12**, 95 (1965).

and φ as calculated by Damany for various values of n and k. The values of n and k were taken from published values referring to various materials for a wavelength of 584 Å [1,18]. It can be seen from the table that the principal angle $\bar{\theta}_i$ and the polarizing angle φ are always quite similar.

9.3   THE OPTICAL CONSTANTS n AND k

The values of n and k are best determined in the vacuum uv region by a reflectance method first described in detail by Tousey [19] in 1939 and later by Simon [20]. The method consists of measuring the reflectance R of a surface at more than one angle of incidence. If the incident light is partially plane polarized, it is also necessary to measure the degree of polarization P. The total reflectance is given by

$$R = \tfrac{1}{2}R_\perp(1 + P) + \tfrac{1}{2}R_\parallel(1 - P), \tag{9.16}$$

when $P = (I_\perp - I_\parallel)/(I_\perp + I_\parallel)$ and where $R_\perp$ and $R_\parallel$ are the Fresnel coefficients given by (9.9) and (9.10) and are both functions of n, k, and the angle of incidence. Measuring the value of R for at least two angles of incidence, and knowing P, two equations such as (9.16) are obtained. Since the equations cannot be solved explicitly, a graphical solution is employed. Tousey has published curves of constant R as a function of n and k for unpolarized light ($P = 0$) and for several angles of incidence [19]. Figure 9.10 reproduces several of his curves for $\theta_i = 45°$ and $75°$. As an example of their application, consider the reflectance of tungsten measured at these two angles and at a wavelength of 584 Å. R is found to be 22 percent at 45° and 60 percent at 75°. Referring to Fig. 9.10, the reflectance curves for these values intersect at a point where n is equal to 0.66 and k is equal to 0.6. A complete discussion of this technique and the errors involved has been given by Hunter [21].

Another method for determining the index of refraction has been developed by Hunter [22]. However it applies only when $n < 1$ and for a transparent material or one which has a very small value of k. The method involves measuring the reflectance of a material as a function of the angle of incidence, then determining the angle θ_c for which the slope of the reflectance curve is a maximum. The value of n is then given by

$$n = \sin \theta_c. \tag{9.17}$$

For $k = 0$, this angle is the critical angle. Applying this technique to aluminum at a wavelength of 584 Å ($k \sim 0.1$), he obtained a value of $n = 0.72$.

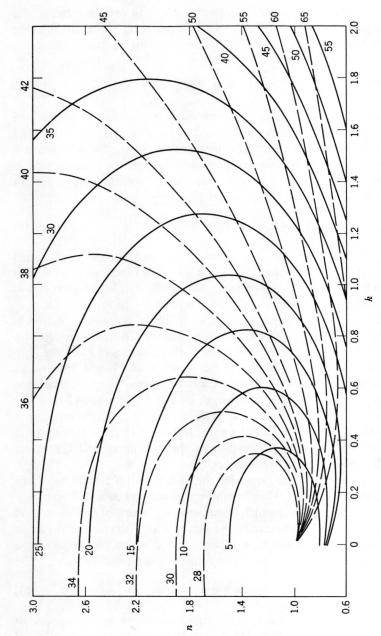

Fig. 9.10 Curves of constant reflectance as a function of *n* and *k*. The solid lines and the innermost numbers represent the reflectance at an angle of incidence of 45°, while the dashed lines represent the reflectance at an angle of incidence of 75° (courtesy R. Tousey [19]).

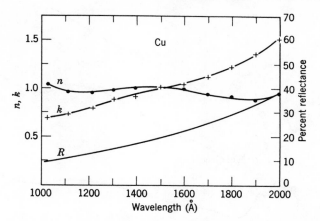

Fig. 9.11 Optical constants and normal incidence reflectance of evaporated copper as a function of wavelength from 1000 to 2000 Å (courtesy L. R. Canfield and G. Hass [23]).

The values of n and k as a function of wavelength for Cu, Ag, Au, Al, In, and ZnS are shown in Figs. 9.11 to 9.16 [23–26]. Table 9.3 lists the values of n and k for MgF_2 [1], SiO [1], ZnS [1], and Al_2O_3 [27]. The optical constants for many other materials have been measured recently and are reported in the literature [28–34]. Because the reflectances of materials in the vacuum uv are very sensitive to surface conditions the values of n and k will also depend on the surface conditions. Aluminum, for example,

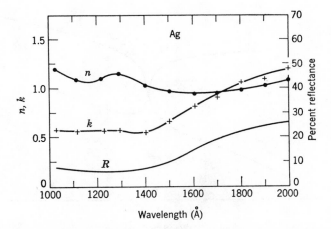

Fig. 9.12 Optical constants and normal incidence reflectance of evaporated silver as a function of wavelength from 1000 to 2000 Å (courtesy L. R. Canfield and G. Hass [23]).

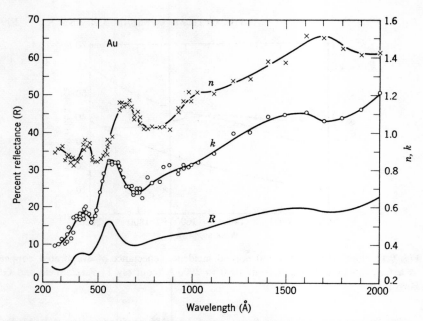

Fig. 9.13 Optical constants and normal incidence reflection of gold in the extreme ultraviolet (courtesy L. R. Canfield, G. Hass, and W. R. Hunter [24]).

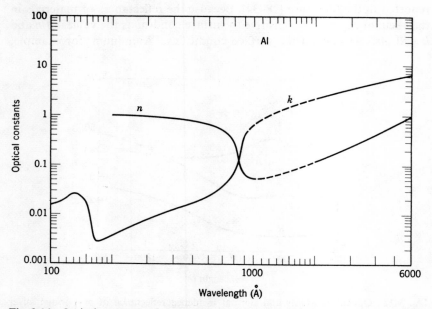

Fig. 9.14 Optical constants of aluminum. The dashed lines represent estimated values of n and k (courtesy W. R. Hunter [25]).

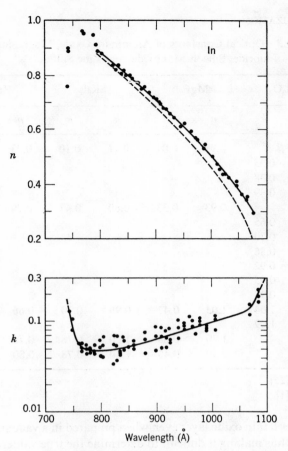

Fig. 9.15 Optical constants of indium from 744 to 1085 Å (courtesy W. R. Hunter [25]).

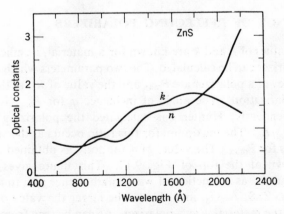

Fig. 9.16 Optical constants of zinc sulfide vacuum deposited onto glass at room temperature (courtesy J. T. Cox, J. E. Waylonis, and W. R. Hunter [26]).

Table 9.3 Optical Constants of Aluminum Oxide, Magnesium
Fluoride, Silicon Monoxide, and Zinc Sulfide

	Al_2O_3[a]		MgF_2[b]		SiO[b]		ZnS[b]	
λ (Å)	n	k	n	k	n	k	n	k
304			0.93	0.07	0.91	0.10	0.93	0.07
507	0.68	0.35						
525	0.66	0.38						
554	0.66	0.49						
584			0.93	0.33	0.80	0.47	0.79	0.20
608	0.66	0.65						
630	0.69	0.70						
740	0.80	0.86						
762	0.92	0.92						
790	0.96	0.95						
835	1.06	0.98						
920	1.20	1.04	1.03	0.42	0.96	0.70	0.66	0.60
1032	1.38	1.09						
1048			1.20	0.68	1.20	0.78	0.76	0.75
1216			1.30	0.45	1.30	0.78	0.80	1.00

[a] See reference [25].
[b] See reference [1].

rapidly acquires a thin oxide layer even when prepared in a vacuum of 10^{-5} torr or better, thus making it difficult to determine the true values of n and k for the pure metal.

9.4   EFFICIENCY OF REFLECTING POLARIZERS

When the values of n and k are known for a material, its efficiency as a reflecting polarizer can be calculated. The two parameters which characterize the efficiency of a polarizer are P_{max} and the value of $R_\perp$ at the angle of maximum polarization. The angle of incidence φ for P_{max} is given by (9.15). Independently, Hunter has calculated the polarizing angle by maximizing $R_\perp/R_\parallel$. The maximum for this ratio occurs at the same angle of incidence as for P_{max}. The values of φ for gold, as obtained by Hunter [35], are shown at the top of Fig. 9.17. This figure gives the ratio $(R_\perp/R_\parallel)_{max}$ and $R_\perp$ as a function of wavelength (hence as a function of n and k) for Au, ZnS, Al_2O_3, and glass. The larger the value of n/k, the more effective the material is as a polarizer. As can be seen from Fig. 9.17, gold has the most uniform polarization characteristics with $(R_\perp/R_\parallel)_{max}$

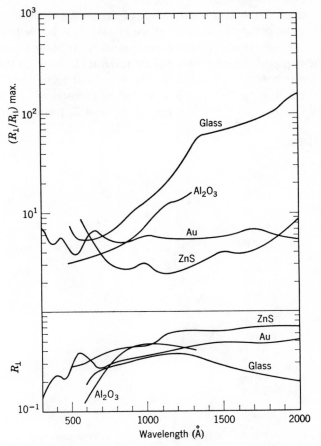

Fig. 9.17 $(R_\perp/R_\parallel)_{max}$ and $R_\perp$ calculated for glass, Au, Al_2O_3, and ZnS (courtesy W. R. Hunter [35]).

varying between 5 and 7 from 2000 to 600 Å. The reflectance of gold varies little with age [24] and maintains a reasonable value of R at the polarizing angle; thus, it is a good material to use for a reflecting polarizer.

Rabinovitch et al. [36] have discussed a technique to determine the degree of polarization produced by a grating monochromator. Their method consists of simply measuring the reflectance of a mirror at an angle of incidence of 45° when the mirror is oriented such that the plane of incidence is (a) perpendicular to the length of the exit slit of the monochromator and (b) parallel to the exit slit. An analysis of this method is given below. Hamm et al. [37] have also derived a technique for measuring the degree of polarization produced by a monochromator. Their method requires measuring the intensity of the reflected radiation from two mirrors in series. The planes of incidence of the two mirrors are rotated with respect to the plane of incidence of the grating, that is, the horizontal plane. Three measurements of the reflected intensity are made (a) when the planes of incidence of the mirrors are both horizontal, (b) when the plane of the first one is horizontal and that of the other is vertical, and (c) when the plane of the first one is vertical and that of the other is horizontal. This method can also provide the values of $R_\perp$ and $R_\parallel$ for the mirrors as

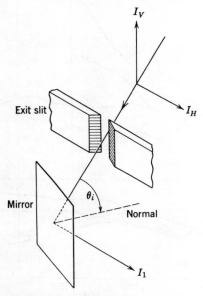

Fig. 9.18 Mirror analyzer for determining the degree of polarization produced by a monochromator.

well as determining the degree of polarization of the mirrors at the specific angles of incidence used. The analysis given below shows how the above parameters can be determined using a single mirror once the degree of polarization of the monochromator is known. The formulas derived, although in different form, are similar to those given by the above authors.

Consider partially plane-polarized radiation from any source, including that from a grating monochromator, with the maximum of the electric vector either in the vertical or horizontal plane. Let the intensity of the component of radiation vibrating in the vertical plane be I_V, and in the horizontal plane, I_H (see Fig. 9.18). Defining the degree of polarization of this radiation as $P = (I_V - I_H)/(I_V + I_H)$, we have from (9.16) for the reflectances R_1 and R_2 measured when the planes of incidence are horizontal and vertical, respectively,

$$R_1 = \tfrac{1}{2}R_\perp(1 + P) + \tfrac{1}{2}R_\parallel(1 - P) \tag{9.18}$$

and

$$R_2 = \tfrac{1}{2}R_\perp(1 - P) + \tfrac{1}{2}R_\parallel(1 + P). \tag{9.19}$$

Solving these two linear equations for $R_\perp$ and $R_\parallel$, we get

$$R_\perp = \frac{\tfrac{1}{2}(R_1 + R_2) + (R_1 - R_2)}{2P}. \tag{9.20}$$

and

$$R_\parallel = \frac{\tfrac{1}{2}(R_1 + R_2) + (R_2 - R_1)}{2P}. \tag{9.21}$$

Thus when the degree of polarization of the incident radiation is known, $R_\perp$ and $R_\parallel$ can be determined as a function of the angle of incidence θ_i by simply measuring R_1 and R_2 as a function of θ_i. If P is unknown, it can be determined by using the relation discovered by Abeles [38], which is valid only for θ_i equal to 45°, namely,

$$R_\parallel = R_\perp{}^2 \qquad (\theta_i = 45°). \tag{9.22}$$

This relation is strictly true only for reflectances at the interface of two isotropic, homogeneous media which have the same magnetic permeability. Using (9.22) to eliminate $R_\perp$ and $R_\parallel$ in (9.20) and (9.21) and solving for P, we get

$$P = \frac{R_2 - R_1}{1 + R_1 + R_2 - [1 + 4(R_1 + R_2)]^{1/2}} \qquad (\theta_i = 45°). \tag{9.23}$$

We can also determine the ability of the mirror to polarize radiation by using (9.20) and (9.21). Let $\rho(\theta_i) = (R_\perp - R_\parallel)/(R_\perp + R_\parallel)$ be the degree of polarization of the radiation reflected from the mirror. Let I_1 be the intensity of the reflected signal measured when the plane of incidence is horizontal, and I_2 when it is vertical. Then $R_1 = I_1/(I_V + I_H)$ and $R_2 = I_2/(I_V + I_H)$. Thus

$$\rho(\theta_i) = \frac{1}{P}\left(\frac{I_1 - I_2}{I_1 + I_2}\right). \tag{9.24}$$

Once ρ is known, P can be found for any source.

No appreciable polarization of radiation has been reported using normal incidence monochromators ($\theta_i < 10°$). However, using the Seya mounting ($\theta_i \sim 35°$), the emerging monochromatic radiation is considerably polarized. Figure 9.19 shows the degree of polarization in terms of the parameter $g = I_V/I_H$, where $P = (g - 1)/(g + 1)$, as determined by Rabinovitch et al. [36] for a 0.5 m Seya monochromator. Two

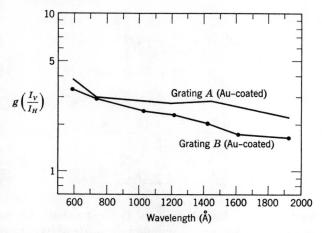

Fig. 9.19 Values of $g(I_V/I_H)$ as a function of wavelength for two gold coated gratings. Grating A was blazed for 700 Å at normal incidence and grating B was blazed for 1500 Å (courtesy K. Rabinovitch, L. R. Canfield, and R. P. Madden [36]).

gratings were examined, both replicas with 1200 lines per mm and coated with gold. Grating A was blazed for 700 Å, while grating B was blazed for 1500 Å. Both gratings showed considerable polarization of the incident radiation, the degree of polarization increasing towards shorter wavelengths. They further showed that the type of coating on the grating was a major influence on the degree of polarization. The characteristics of grating B were completely changed when this grating was overcoated with aluminum. Figure 9.20 shows the data obtained by Hamm et al. [37] for

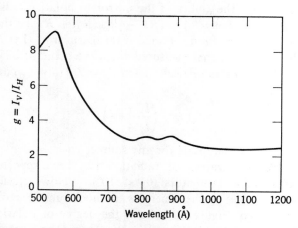

Fig. 9.20 Values of $g(I_V/I_H)$ as a function of wavelength for a platinized grating blazed for 700 Å (courtesy R. N. Hamm, R. A. MacRae, and E. T. Arakawa [37]).

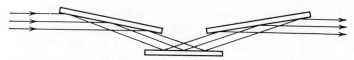

Fig. 9.21 A polarizer using three mirrors to preserve the direction of the incident radiation. Radiation is incident at angles of 80°, 70°, and 80°, respectively.

a platinized replica blazed at 700 Å and used in a 0.5 m Seya monochromator. At 550 Å, P is 80 percent. Hamm et al. also measured the value of g for plane mirrors of gold and silver. They found for an angle of incidence of 45° that g varied between 4 and 5 for gold and between 5 and 7.7 for silver. At $\theta_i = 45°$, therefore, a silver mirror is a more efficient polarizer than gold. To maintain the direction of radiation using a reflecting polarizer, Hamm et al. positioned three gold mirrors as shown in Fig. 9.21 with the radiation incident at angles of 80°, 70°, and 80°, respectively. The degree of polarization, expressed in terms of g, is shown in Fig. 9.22 as a function of wavelength between 500 and 1200 Å. The transmission of such a system was about four percent of the unpolarized light.

Metcalf and Baird [39] have shown that it is possible to produce circularly polarized vacuum uv radiation by applying pressure to the edges of a plate of cleaved LiF about 2 to 4 mm thick. The unstrained LiF crystal will transmit unaltered plane polarized radiation; however, as the applied pressure is increased, the crystal becomes optically active and converts the incident plane polarized radiation into elliptically polarized

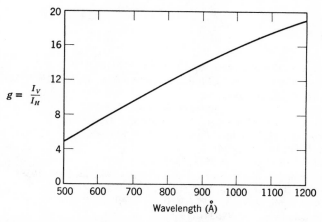

Fig. 9.22 Values of $g(I_V/I_H)$ as a function of wavelength for the gold-coated reflecting polarizer shown in Fig. 9.21 (courtesy R. N. Hamm, R. A. MacRae, and E. T. Arakawa [37]).

radiation. That is, the stressed LiF crystal acts as a retardation plate. The degree of retardation increases linearly with the applied pressure until the retardation is 90° at which point the radiation transmitted is circularly polarized. The pressure necessary to produce circularly polarized radiation can be determined using an arrangement such as that shown in Fig. 9.23. The transmitted radiation is analyzed by a LiF plate inclined at a constant angle of 30° to the direction of the incident beam. The analyzer is then rotated about the incident beam, while the analyzed radiation is

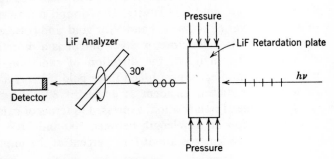

Fig. 9.23 LiF retardation plate and analyzer for the production and detection of circularly polarized radiation.

observed by the detector. If the retardation plate is producing elliptically polarized radiation, the signal from the detector will oscillate in phase with the rotating analyzer. When circularly polarized radiation is produced, the signal from the detector will be steady.

REFERENCES

[1] T. T. Cole and F. Oppenheimer, *Appl. Optics* **1**, 709 (1962).

[2] G. Scheibe, St. Hartwig, and R. Müller, *Z. Elektrochem.* **49**, 373 (1943).

[3] R. Müller, *Optik* **20**, 510 (1963).

[4] K. S. Krishnan and A. C. Dasgupta, *Nature* **126**, 12 (1930).

[5] K. S. Krishnan and B. Mukhopadhyay, *Nature* **132**, 411 (1933).

[6] F. A. Jenkins and H. E. White, *Fundamentals of Physical Optics* (McGraw-Hill, New York, 1937), 1st ed., p. 341.

[7] W. C. Johnson, *Rev. Sci. Instr.* **35**, 1375 (1964).

[8] A. Duncanson and R. W. H. Stevenson, *Proc. Phys. Soc.* **72**, 1001 (1958).

[9] J. L. Weinberg, *Appl. Optics.* **3**, 1057 (1964).

[10] W. C. Walker, Appl. *Optics* **3**, 1457 (1964).

[11] E. G. Schneider, *Phys. Rev.* **49**, 341 (1936).

[12] W. Konig, *Handbuch der Physik* (Springer-Verlag, Berlin, 1928) Vol. 20, p. 242, and quoted by R. Tousey, *J. Opt. Soc. Am.* **29**, 235 (1939).

[13] See reference 6, p. 400.

[14] M. Born and E. Wolf, *Principles of Optics* (Macmillan, New York, 1964) 2nd. ed, p. 620. NB. k is identical to n_κ used in this reference.

[15] S. P. F. Humphreys-Owen, *Proc. Phys. Soc. London* **77**, 949 (1961).

[16] H. Damany, *J. Opt. Soc. Am.* **55**, 1558 (1965).

[17] H. Damany, *Optica Acta.* **12**, 95 (1965).

[18] R. P. Madden, L. R. Canfield, and G. Hass, *J. Opt. Soc. Am.* **53**, 620 (1963).

[19] R. Tousey, *J. Opt. Soc. Am.* **29**, 235 (1939).

[20] I. Simon, *J. Opt. Soc. Am.* **41**, 336 (1951).

[21] W. R. Hunter, *J. Opt. Soc. Am.* **55**, 1197 (1965).

[22] W. R. Hunter, *J. Opt. Soc. Am.* **34**, 1565 (1963).

[23] L. R. Canfield and G. Hass, *J. Opt. Soc. Am.* **55**, 61 (1965).

[24] L. R. Canfield, G. Hass, and W. Hunter, *J. de Physique* **25**, 124 (1964).

[25] W. R. Hunter, *J. Opt. Soc. Am.* **54**, 208 (1964); *J. de Physique* **25**, 154 (1964).

[26] J. T. Cox, J. E. Waylonis, and W. R. Hunter, *J. Opt. Soc. Am.* **49**, 807 (1959).

[27] G. H. C. Freeman, Brit. *J. Appl. Phys.* **16**, 927 (1965).

[28] J. G. Carter, R. H. Huebner, R. N. Hamm, and R. D. Birkhoff, *Phys. Rev.* **137**, A639 (1965), Graphite.

[29] H. R. Phillip and E. A. Taft, *Phys. Rev.* **127**, 159 (1962), Diamond.

[30] W. C. Walker and J. Osantowski, *Phys. Rev.* **134**, A153 (1964), Diamond.

[31] R. Kato, *J. Phys. Soc. Japan* **16**, 2525 (1961), Lithium fluoride.

[32] O. P. Rustgi, J. S. Nodvik, and G. L. Weissler, *Phys. Rev.* **122**, 1131 (1961), Germanium.

[33] R. Tousey, *Phys. Rev.* **50**, 1057 (1936), Calcium fluoride.

[34] T. Sasaki, H. Fukutani, and K. Ishiguro, *J. Appl. Phys. Japan* **4**, Suppl. 1, 527 (1965), Glass and quartz.

[35] W. R. Hunter, *J. Appl. Phys. Japan* **4**, Suppl. 1, 520 (1965).

[36] K. Rabinovitch, L. R. Canfield, and R. P. Madden, *Appl. Optics* **4**, 1005 (1965).

[37] R. N. Hamm, R. A. MacRae, and E. T. Arakawa, *J. Opt. Soc. Am.* **55**, 1460 (1965).

[38] F. Abelès, *Compt. Rend.* **230**, 1942 (1950).

[39] H. Metcalf and J. C. Baird, *Appl. Optics* **5**, 1407 (1966).

10

Wavelength Standards

In 1960, the Eleventh General Conference on Weights and Measures in Paris adopted an orange-red line of Kr 86 as the standard of wavelength. The meter was then defined as exactly 1,650,763.73 wavelengths of Kr 86 as measured in vacuum. Thus the primary standard of wavelength is 6057.802106 Å, where 1 Å is exactly 10^{-8} cm.

For the practical determination of wavelengths, secondary standards have been measured interferometrically against the primary standard. These secondary standards are obtained from the arc spectra of iron, thorium, the rare gases, and other elements. Unfortunately, these standards exist only for wavelengths greater than 2000 Å. This is because of the practical difficulties encountered with interferometers at shorter wavelengths. Attempts to determine secondary standards below 2000 Å have been made by MacAdam [1]. However, no values have been reported. It may be possible to use reflection echelons down to 1200 Å with the enhanced reflection coatings recently reported by Hass and co-workers [2,3].

Tertiary standards must be adopted in the vacuum uv. Two types exist. The method of overlapping orders in a grating spectrum and the calculated Ritz standards. The method of overlapping orders is a very direct method. The line under investigation is observed in its second, third, or higher order where it falls close to lower order secondary standards. The wavelength of the line is then determined by interpolation. The accuracy of this method, although limited by coincidence errors, should be of the order of 0.01 Å or better.

The use of the Ritz combination principle is the most precise method for establishing vacuum uv standard wavelengths. The method consist of determining the relative values of atomic energy levels from observed secondary standards then calculating the wavelength of the standard vacuum uv line from the difference between the two levels involved in the transition. The technique is illustrated in Fig. 10.1 for the 3d-4f transition of Ar II. The wavelength errors in the vacuum uv Ritz standards are generally smaller than those of the secondary standards from which they

320

were derived. The reason for this is that the errors of the determined energy levels are approximately constant, while the wavelength errors are proportional to the square of the wavelength; that is, $\Delta\lambda = \lambda^2\Delta\sigma$, where σ is the wavenumber. At 3000 Å, an uncertainty of 0.1 cm^{-1} in σ gives

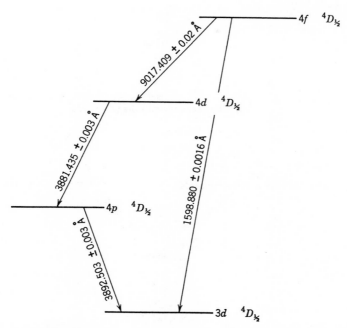

Fig. 10.1 The $3d$–$4f$ transition of Ar II illustrating the principle of Ritz standards. The wavelengths listed in the figure are vacuum wavelengths as measured by Johansson [14]. The calculated $3d$–$4f$ transition is in good agreement with that measured by Johansson at 1598.872 Å.

a wavelength uncertainty of 0.01 Å, while at 1000 Å, the same uncertainty in σ gives a $\Delta\lambda$ of 0.001 Å.

Having established vacuum uv standards, the unidentified lines can be determined by interpolation. To proceed to shorter wavelengths, it is necessary to use the interpolated values as "fourth order" standards from which shorter wavelength Ritz standards can be calculated, and so on. Figure 10.2 summarizes the development of vacuum uv standards. Some of the earlier tabulations of vacuum wavelengths were given by Boyce and Robinson [4], Weber and Watson [5], and later by Boyce [6]. More recently, Edlén [7] has compiled and published a selected list of the more intense wavelengths from 2000 to 14 Å. Additional wavelengths have been published by Wilkinson et al. [8], Shenstone [9], Meissner et al. [10], Kaufman and Andrew [11], Radziemski and Andrews [12], and others. A very complete compilation of the Ar II and C I spectrum has been given by Minnhagen [13] and Johansson [14], respectively. Iglesias [15] has reported 68 new Ritz standards for Mn II in the range 1162 to 1935 Å.

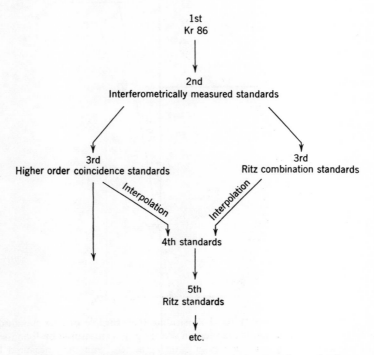

Fig. 10.2 Development of vacuum uv wavelength standards.

The wavelengths of the resonance series of the rare gases provide very useful and commonly used standards and, therefore, have been tabulated in Table 10.1, the values being taken from the work of Herzberg [16], Martin [17], and Petersson [18]. Very few standard wavelengths exist below 500 Å. Table 10.2, extracted from extensive tables published by Garcia and Mack [19], lists the calculated wavelengths for the resonance lines of the isoelectronic sequence H I to Ca XX covering the range 2 to 1215 Å. These lines consist of close doublets, the intensities being in the ratio of 1:2. The doublet separation $\Delta\lambda$ is given in the last row of the table. The wavelengths listed represent the positions of the center of gravity of the doublets. The individual components have the wavelengths $\lambda + 2\Delta\lambda/3$

Table 10.1 Calculated Ritz Standards in the Vacuum Ultraviolet (Å)[a]

He I	Ne I	Ar I	Kr I	Xe I
584.3340	743.718	1066.659	1235.838	1469.610
537.0296	735.895	1048.219	1164.867	1295.586
522.2128	629.738	894.310	1030.023	1250.207
515.6165	626.822	879.946	1003.550	1192.036
512.0982	619.101	876.057	1001.061	1170.410
509.9979	618.671	869.754	963.374	1129.307
508.6431	615.627	866.800	953.404	1110.713
507.7178	602.725	842.805	951.056	1099.716
507.0576	600.036	835.002	946.535	1085.442
506.5702	598.890	834.392	945.441	1078.584
506.2000	598.705	826.364	928.711	1070.409
505.9122	595.919	825.346	923.713	1068.167

[a] Values for He I were obtained from Herzberg [16] and Martin [17], while the data for Ne I to Xe I were obtained from Petersson [18].

Table 10.2 Calculated Wavelengths for the Resonance Lines of One-Electron Spectra from H I to Ca XX[a]

	$1s$–$2p$	$1s$–$3p$	$1s$–$4p$	$1s$–$5p$	$1s$–$6p$	$1s$–$7p$
H I	1215.6701	1025.7223	972.5368	949.7430	937.8035	930.7483
D I	1215.3394	1024.4433	972.2722	949.4847	937.5484	930.4951
He II	303.7822	256.3170	243.0266	237.3307	234.3472	232.5842
Li III	134.9976	113.9050	107.9989	105.4678	104.1420	103.3585
Be IV	75.9277	64.0648	60.7431	59.3195	58.5739	58.1333
B V	48.5873	40.9964	38.8708	37.9599	37.4827	37.2008
C VI	33.7360	28.4656	26.9898	26.3573	26.0260	25.8303
N VII	24.7810	20.9098	19.8258	19.3613	19.1179	18.9741
O VIII	18.9689	16.0059	15.1762	14.8206	14.6343	14.5243
F IX	14.9841	12.6438	11.9884	11.7075	11.5604	11.4734
Ne X	12.1339	10.2389	9.7082	9.4807	9.3616	9.2912
Na XI	10.0250	8.4595	8.0211	7.8332	7.7348	7.6766
Mg XII	8.4210	7.1062	6.7379	6.5801	6.4974	6.4486
Al XIII	7.1727	6.0529	5.7393	5.6049	5.5345	5.4929
Si XIV	6.1822	5.2172	4.9469	4.8311	4.7704	4.7346
P XV	5.3831	4.5430	4.3076	4.2068	4.1540	4.1228
S XVI	4.7291	3.9912	3.7845	3.6959	3.6495	3.6221
Cl XVII	4.1871	3.5339	3.3508	3.2724	3.2313	3.2071
Ar XVIII	3.7329	3.1506	2.9875	2.9176	2.8810	2.8593
K XIX	3.3485	2.8263	2.6799	2.6172	2.5844	2.5650
Ca XX	3.0203	2.5493	2.4174	2.3608	2.3312	2.3137
$\Delta\lambda$ =	0.00540	0.00114	0.00043	0.00021	0.00012	0.00007

[a] Extracted from J. D. Garcia and J. E. Mack, *J. Opt. Soc. Am.* **55,** 654 (1965).

and $\lambda - \Delta\lambda/3$. The authoritative source for wavelength standards is Commission 14 of the International Astronomical Union [20].

To aid in the identification of the origin and wavelength of a line in a spectrum, several tables of wavelengths of the elements have been published, notably by Boyce and Robinson [4], Kayser and Ritschl [21], Moore [22], and Kelly [23]. The compilation by Kelly is the largest and most recent. Edlén [7] has tabulated the strongest lines for different ionization stages of some light elements commonly found as impurity lines in many spectra. This list is reproduced here in Table 10.3 by kind permission from Professor Edlén. The particular advantage of the lines given in Table 10.3 is that they are generally among the first lines to appear in a spectrum. If they do not appear or are very faint, other lines of the same spectrum should not be present. However, if these lines are strong, other members of the same spectrum are likely to be present and reference to a more complete finding list is necessary. The figures given in parentheses after the wavelengths represent intensities on a linear scale, normalized to 20 or 30 for the strongest lines of each spectrum. For example, the intensity of an O III line is given relative to other O III lines and is not compared to N III or O I lines, etc. The wavelengths of the Si I, Si II, Al I, N I, and C I lines appearing in Table 10.3 have been revised in accord with the recent work of Kaufman et al. [24] (Si I), Kaufman and Ward [25] (Si II, N I), Eriksson and Isberg [26] (Al I), and Johansson [14] (C I).

The determination of the wavelength of an unknown line is made by measuring the displacement in millimeters of the line from an auxiliary standard by means of a comparator. Knowing the plate factor (reciprocal dispersion) in angstroms per millimeter, the displacement in angstroms can be found. Although the plate factor can be determined from the equation $d\lambda/dl = d\cos\beta/nR$ (see Chapter 2), it is more accurate to determine the plate factor from the separation of two standard lines in the vicinity of the measurements. The use of comparators and the practical details of measuring the position of spectral lines has been discussed by Sawyer [27] and by Harrison et al. [28].

The most commonly used sources for the production of vacuum uv spectral lines suitable for wavelength standards have been the hollow cathode and the electrodeless discharge. The hollow cathode can be constructed with the material under investigation, or crystals of the element can be placed within the cathode. The carrier gas is usually helium at a pressure of 1 to 4 torr. As mentioned earlier in Chapter 5, the hollow cathode discharge produces both the first and second spectra of the material investigated. The electrodeless discharge can be conveniently powered using microwave frequencies. Figure 10.3 shows the arrangement of a microwave source (125 watt) for the production of Si and Ge lines,

Table 10.3 Condensed Finding List of Vacuum Ultraviolet
Atomic Lines of Light Elements[a]

1990.530	(5)	Al II	1656.929	(4)	C I	1306.029	(10)	O I
1988.993	(6)	Si I	1656.267	(5)	C I	1304.858	(20)	O I
1935.88	(5)	Al III	1640.532			1303.323	(7)	Si III
1930.905	(15)	C I	1640.332	(5)	He II	1302.168	(30)	O I
1901.337	(10)	Si I	1624.37	(5)	B II	1301.149	(5)	Si III
1862.795	(15)	Al IHI	1624.16	(4)	B II	1298.946	(20)	Si III
1862.318	(10)	Al II	1623.99	(18)	B II	1298.892	(4)	Si III
1858.031	(6)	Al II	1623.77	(4)	B II	1296.726	(5)	Si III
1855.928	(2)	Al II	1623.57	(5)	B II	1294.545	(7)	Si III
1854.720	(30)	Al III	1611.85	(15)	Al III	1265.001	(3)	Si II
1845.472	(5)	Si I	1605.75	(8)	Al III	1264.737	(30)	Si II
1850.672	(10)	Si I	1561.438	(20)	C I	1260.421	(15)	Si II
1847.473	(7)	Si I	1561.340	(4)	C I	1251.164	(4)	Si II
1845.520	(5)	Si I	1560.709	(4)	C I	1248.426	(3)	Si II
1826.400	(10)	B I	1560.682	(10)	C I	1247.383	(3)	C III
1825.899	(5)	B I	1560.310	(5)	C I	1246.738	(2)	Si II
1816.928	(10)	Si II	1550.77	(15)	C IV	1242.79	(15)	N V
1808.013	(5)	Si II	1548.20	(30)	C IV	1238.80	(30)	N V
1776.307	(6)	Be II	1533.432	(20)	Si II	1217.643	(4)	O I
1776.100	(3)	Be II	1526.707	(10)	Si II	1215.670	(30)	H I
1769.140	(4)	Al I	1512.419	(1)	Be II	1215.185		
1767.735	(3)	Al II	1512.407	(9)	Be II	1215.088	(3)	He II
1766.385	(4)	Al I	1512.269	(5)	Be II	1206.555	(10)	Si III
1765.811	(2)	Al II	1494.675	(10)	N I	1206.500	(30)	Si III
1765.636	(4)	Al I	1492.820	(2)	N I	1200.710	(10)	N I
1763.947	(6)	Al II	1492.625	(20)	N I	1200.223	(20)	N I
1763.874	(1)	Al II	1491.765	(2)	Be I	1199.550	(30)	N I
1762.899	(2)	Al I	1417.237	(3)	Si III	1194.500	(3)	Si II
1761.979	(2)	Al II	1402.770	(15)	Si IV	1193.289	(2)	Si II
1760.103	(3)	Al II	1393.755	(30)	Si IV	1176.370	(5)	C III
1745.252	(8)	N I	1384.14	(6)	Al III	1175.987	(4)	C III
1742.729	(15)	N I	1379.67	(3)	Al III	1175.711	(15)	C III
1724.97	(20)	Al II	1371.29	(5)	O V	1175.590	(3)	C III
1721.26	(12)	Al II	1362.460	(30)	B II	1175.263	(4)	C III
1719.43	(5)	Al II	1335.708	(20)	C II	1174.933	(5)	C III
1718.52	(5)	N IV	1335.663	(2)	C II	1152.152	(8)	O I
1673.405	(5)	Ar III	1334.532	(10)	C II	1134.981	(30)	N I
1670.786	(30)	Al II	1329.600	(5)	C I	1134.415	(20)	N I
1669.666	(8)	Ar III	1329.577	(15)	C I	1134.166	(10)	N I
1661.478	(8)	Be I	1329.123	(3)	C I	1128.340	(9)	Si IV
1658.121	(5)	C I	1329.100	(5)	C I	1128.325	(1)	Si IV
1657.907	(4)	C I	1329.085	(4)	C I	1122.485	(5)	Si IV
1657.379	(3)	C I	1328.834	(4)	C I	1113.228	(20)	Si III
1657.008	(15)	C I	1312.591	(3)	Si III	1109.965	(12)	Si III

Table 10.3 (*continued*)

1108.368	(4)	Si III	903.962	(6)	C II	765.140	(30)	N IV
1085.701	(30)	N II	903.624	(3)	C II	764.357	(10)	N III
1085.546	(5)	N II	901.162	(9)	Ar IV	763.340	(5)	N III
1084.580	(15)	N II	900.362	(5)	Ar IV	762.001	(5)	O V
1084.562	(5)	N II	894.310	(5)	Ar I	761.130	(4)	O V
1083.990	(7)	N II	887.404	(10)	Ar III	760.445	(15)	O V
1066.659	(20)	Ar I	883.179	(8)	Ar III	760.229	(3)	O V
1066.629	(5)	Si IV	879.622	(6)	Ar III	759.440	(4)	O V
1048.219	(25)	Ar I	878.728	(20)	Ar III	758.677	(5)	O V
1037.62	(15)	O VI	876.057	(4)	Ar I	745.322	(7)	Ar II
1037.018	(15)	C II	875.534	(8)	Ar III	744.925	(8)	Ar II
1036.337	(8)	C II	871.099	(10)	Ar III	743.718	(12)	Ne I
1036.319	(5)	Be II	866.800	(4)	Ar I	740.270	(10)	Ar II
1036.299	(10)	Be II	858.559	(4)	C II	735.895	(30)	Ne I
1031.93	(30)	O VI	858.092	(2)	C II	730.929	(5)	Ar II
1028.157	(2)	O I	850.602	(25)	Ar IV	723.361	(5)	Ar II
1027.431	(4)	O I	843.772	(20)	Ar IV	718.562	(20)	O II
1025.762	(6)	O I	840.029	(15)	Ar IV	718.484	(30)	O II
1025.722	(10)	H I	835.292	(20)	O III	715.65	(3)	Ar V
1010.374	(8)	C II	835.096	(4)	O III	715.60	(4)	Ar V
1010.092	(5)	C II	834.88	(4)	Ar V	713.81	(10)	Ar VIII
1009.862	(3)	C II	834.467	(20)	O II	709.197	(5)	Ar V
997.386	(10)	Si III	833.742	(15)	O III	703.850	(30)	O III
994.790	(6)	Si III	833.332	(15)	O II	702.899	(15)	O III
993.519	(2)	Si III	832.927	(5)	O III	702.822	(10)	O III
991.579	(30)	N III	832.762	(8)	O II	702.332	(10)	O III
991.514	(3)	N III	827.052	(5)	Ar V	700.277	(8)	Ar IV
989.790	(15)	N III	822.161	(4)	Ar V	700.24	(20)	Ar VIII
977.020	(30)	C III	818.128	(6)	Si IV	699.408	(4)	Ar IV
972.537	(5)	H I	815.053	(3)	Si IV	696.212	(2)	Al III
955.335	(3)	N IV	801.913	(5)	Ar IV	695.817	(4)	Al III
932.053	(15)	Ar II	801.409	(15)	Ar IV	693.952	(6)	B II
924.274	(5)	N IV	801.086	(15)	Ar IV	689.007	(12)	Ar IV
923.669	(4)	N IV	800.573	(5)	Ar IV	687.345	(10)	C II
923.211	(15)	N IV	796.661	(2)	O II	687.053	(5)	C II
932.045	(3)	N IV	790.203	(30)	O IV	686.335	(6)	N III
922.507	(4)	N IV	790.103	(3)	O IV	685.816	(30)	N III
921.982	(5)	N IV	787.710	(15)	O IV	685.513	(12)	N III
919.782	(25)	Ar II	780.324	(10)	Ne VIII	684.996	(6)	N III
916.703	(30)	N II	775.965	(20)	N II	683.287	(8)	Ar IV
916.015	(8)	N II	772.385	(3)	N III	679.400	(6)	Ar II
915.962	(4)	N II	771.901	(2)	N III	677.951	(5)	Ar II
915.612	(4)	N II	771.544	(1)	N III	677.147	(20)	B III
904.480	(3)	C II	770.409	(20)	Ne VIII	677.004	(10)	B III
904.142	(15)	C II	769.152	(15)	Ar III	676.241	(6)	Ar II

326

Table 10.3 (*continued*)

671.852 (6) Ar II	558.481 (5) Ar V	469.865 (15) Ne IV
670.948 (5) Ar II	555.639 (4) Ar VI	469.817 (15) Ne IV
666.010 (6) Ar II	555.262 (6) O IV	465.221 (30) Ne VII
661.869 (5) Ar II	554.514 (30) O IV	463.938 (7) Ar V
660.286 (5) N II	554.074 (12) O IV	463.263 (15) Na V
645.178 (5) N II	553.328 (6) O IV	462.388 (20) Ne II
644.837 (3) N II	551.371 (8) Ar VI	462.007 (25) Ar VI
644.634 (1) N II	548.905 (5) Ar VI	461.051 (10) Na V
644.148 (5) O II	544.731 (4) Ar VI	460.725 (30) Ne II
643.256 (5) Ar III	543.891 (20) Ne IV	459.897 (5) Na V
641.808 (12) Ar III	542.073 (15) Ne IV	459.633 (10) C III
637.282 (20) Ar III	541.127 (8) Ne IV	459.521 (6) C III
629.732 (30) O V	538.312 (5) C III	459.462 (2) C III
629.738 (6) Ne I	538.149 (3) C III	458.155 (2) Si IV
626.822 (7) Ne I	538.080 (1) C III	457.815 (4) Si IV
625.852 (6) O IV	537.030 (12) He I	457.475 (20) Ar VI
625.130 (4) O IV	527.693 (6) Ar V	455.270 (5) Ne II
624.617 (2) O IV	525.795 (30) O III	454.072 (6) Ne VI
619.101 (4) Ne I	524.189 (5) Ar V	452.745 (4) Ne VI
618.671 (5) Ne I	522.213 (5) He I	451.843 (2) Ne VI
615.627 (5) Ne I	518.271 (5) B III	449.065 (18) Ar V
609.829 (6) O IV	518.244 (10) B III	447.813 (10) Ne II
608.395 (3) O IV	508.182 (30) O III	446.949 (8) Ar V
599.598 (30) O III	507.683 (20) O III	446.591 (5) Ne II
597.818 (5) O III	507.391 (10) O III	446.252 (10) Ne II
596.694 (4) Ar VI	494.382 (20) Na VI	445.997 (5) Ar V
588.921 (5) Ar VI	494.160 (4) Na VI	445.032 (7) Ne II
585.754 (15) Ar VII	491.950 (20) Na VII	435.649 (10) Ne VI
584.334 (30) He I	491.340 (12) Na VI	433.176 (5) Ne VI
574.281 (5) C III	491.240 (4) Na VI	421.609 (15) Ne IV
572.336 (20) Ne V	491.050 (8) Ne III	419.714 (6) C IV
572.106 (4) Ne V	490.310 (7) Ne III	419.525 (3) C IV
569.830 (12) Ne V	489.641 (4) Ne III	416.198 (20) Ne V
569.759 (4) Ne V	489.580 (6) Na VI	412.240 (8) Na IV
568.418 (5) Ne V	489.501 (15) Ne III	411.333 (7) Na IV
566.613 (2) Si III	488.868 (7) Ne III	410.540 (6) Na IV
564.529 (5) Ne VII	488.103 (8) Ne III	410.371 (15) Na IV
562.992 (4) Ne VII	486.740 (10) Na VII	409.615 (7) Na IV
562.805 (20) Ne VI	482.987 (20) Ne V	408.682 (8) Na IV
562.735 (2) Ne VI	481.361 (10) Ne V	407.136 (10) Ne II
561.728 (15) Ne VII	481.281 (6) Ne V	405.852 (15) Ne II
561.378 (3) Ne VII	480.406 (10) Ne V	403.262 (6) Ne VI
559.947 (4) Ne VII	479.379 (12) Ar VII	401.939 (30) Ne VI
558.61 (5) Ne VII	475.656 (8) Ar VII	401.138 (12) Ne VI
558.595 (10) Ne VI	473.938 (4) Ar VII	400.722 (10) Na V

Table 10.3   (*continued*)

399.820	(6)	Ne VI	309.852	(6)	Al VI	192.906	(10)	O V
395.558	(5)	O III	309.596	(15)	Al VI	192.799	(6)	O V
388.218	(15)	Ne IV	308.560	(7)	Al VI	192.751	(2)	O V
387.141	(10)	Ne IV	308.264	(12)	Na V	184.117	(10)	O VI
386.203	(8)	C III	307.248	(8)	Al VI	183.937	(5)	O VI
384.178	(20)	C IV	307.152	(6)	Na V	173.082	(20)	O VI
384.032	(10)	C IV	303.782	(30)	He II	172.935	(10)	O VI
380.107	(15)	Na III	283.579	(10)	N IV	172.169	(5)	O V
379.308	(8)	Ne III	283.470	(6)	N IV	161.686	(15)	Al IV
378.143	(30)	Na III	283.420	(2)	N IV	160.073	(25)	Al IV
376.375	(10)	Na II	281.397	(15)	Al V	150.124	(5)	O VI
374.441	(4)	N III	278.699	(30)	Al V	150.089	(10)	O VI
374.204	(2)	N III	278.445	(8)	Si VII	118.968	(15)	Si V
372.069	(25)	Na II	276.839	(7)	Si VII	117.860	(20)	Si V
365.594	(20)	Ne V	275.665	(6)	Si VII	100.254	(30)	Be III
361.250	(25)	Na VI	275.352	(15)	Si VII	88.314	(10)	Be III
359.385	(15)	Ne V	274.175	(7)	Si VII	84.758	(3)	Be III
358.72	(20)	Ne IV	272.641	(8)	Si VII	75.928	(30)	Be IV
358.472	(10)	Ne V	266.375	(6)	N V	60.313	(30)	B IV
357.955	(5)	Ne V	266.192	(3)	N V	52.682	(10)	B IV
357.831	(15)	Ne IV	256.317	(10)	He II	50.435	(3)	B IV
356.885	(15)	Al VII	250.940	(7)	Ar VII	48.587	(30)	B V
354.950	(5)	Na VII	249.886	(5)	Ar VII	40.270	(30)	C V
353.776	(10)	Al VII	249.125	(15)	Si VI	34.973	(10)	C V
353.294	(25)	Na VII	247.710	(20)	N V	33.736	(30)	C VI
352.275	(10)	Na VII	247.563	(10)	N V	33.426	(3)	C V
352.160	(5)	Al VII	247.205	(5)	N IV	28.787	(30)	N VI
350.645	(5)	Na VII	246.001	(30)	Si VI	24.898	(10)	N VI
335.050	(6)	N IV	243.760	(20)	Al VI	24.781	(30)	N VII
320.979	(5)	O III	243.027	(5)	He II	23.771	(3)	N VI
319.638	(20)	Na IV	238.573	(6)	O IV	21.800[b]	(30)	O VII
317.641	(15)	Na VI	238.361	(3)	O IV	21.602	(30)	O VII
312.453	(5)	C IV	220.325	(6)	O V	18.969	(30)	O VIII
312.422	(10)	C IV	217.826	(20)	Si VII	18.627	(10)	O VII
312.241	(8)	Al VI	209.303	(5)	N V	17.768	(3)	O VII
310.908	(7)	Al VI	209.270	(10)	N V			

[a] Courtesy B. Edlén [7].
[b] B. C. Fawcett, A. H. Gabriel, B. B. Jones, N. J. Peacock, *Proc. Phys. Soc.* **84,** 257 (1964).

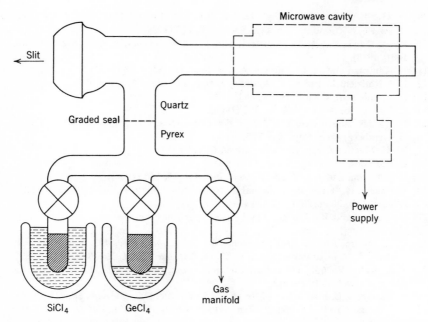

Fig. 10.3   Microwave powered source for the production of vacuum uv wavelength standards (courtesy V. Kaufman et al. [24]).

as used by Kaufman et al. [24]. The vapor pressure of $SiCl_4$ and $GeCl_4$ at dry-ice temperature is about 100 to 200 μ. The vapor from one compound is introduced into the source with a carrier gas of argon ($\sim$40 μ) or helium ($\sim$300 μ) and allowed to flow out through the entrance slit of the monochromator. The widths of the lines produced in this system are less than 0.01 Å.

REFERENCES

[1] D. L. MacAdam, *Phys. Rev.* **50**, 185 (1936).

[2] G. Hass and R. Tousey, *J. Opt. Soc. Am.* **49**, 593 (1959).

[3] P. H. Berning, G. Hass and R. Tousey, *J. Opt. Soc. Am.* **50**, 586 (1960).

[4] J. C. Boyce and H. A. Robinson, *J. Opt. Soc. Am.* **26**, 133 (1936).

[5] R. L. Weber and W. W. Watson, *J. Opt. Soc. Am.* **26**, 307 (1936).

[6] J. C. Boyce, *Revs. Mod. Phys.* **13**, 1 (1941).

[7] B. Edlén, *Rep. Progr. Phys.* (London: Physical Society) **26**, 181 (1963).

[8] P. G. Wilkinson and K. L. Andrew, *J. Opt. Soc. Am.* **53**, 710 (1963); P. G. Wilkinson and D. W. Angel, *J. Opt. Soc. Am.* **52**, 1120 (1962); P. G. Wilkinson, *J. Molecular Spectrosc.* **1**, 288 (1957); **6**, 1 (1961); *J. Opt. Soc. Am.* **47**, 182 (1957).

[9] A. G. Shenstone, *Proc. Roy. Soc. A.* **261,** 153 (1961); **276,** 293 (1963); *Rep. Progr. Phys.* (London: Physical Society) **5,** 210 (1939).

[10] K. W. Meissner, R. D. VanVeld, and P. G. Wilkinson, *J. Opt. Soc. Am.* **48,** 1001 (1958).

[11] V. Kaufman and K. L. Andrew, *J. Opt. Soc. Am.* **52,** 1223 (1962).

[12] L. J. Radziemski and K. L. Andrew, *J. Opt. Soc. Am.* **55,** 474 (1965).

[13] L. Minnhagen, *Arkiv Fysik* **14,** 483 (1958).

[14] L. Johansson, *Arkiv Fysik* **31,** 201 (1966).

[15] L. Iglesias, *Anales Real Soc. Espan. Fis. Quim.* (Madrid), Ser. A **60,** 147 (1964).

[16] G. Herzberg, *Proc. Roy. Soc.* **248,** 309 (1958).

[17] W. C. Martin, *J. Res. Nat. Bur. Stand.* **64A,** 19 (1960).

[18] B. Petersson, *Arkiv Fysik* **27,** 317 (1964).

[19] J. D. Garcia and J. E. Mack, *J. Opt. Soc. Am.* **55,** 654 (1965).

[20] Transactions of the International Astronomical Union **9,** 201 (1951); **10,** 211 (1958); **11B,** 208 (1961); **12A,** 137 (1964).

[21] H. Kayser and R. Ritschl, *Tabelle der Hauptlinien der Spektren aller Elemente* (Springer-Verlag, Berlin, 1939).

[22] C. E. Moore, Atomic Energy Levels, *Nat. Bur. Stand. Circ. 467,* Vols. I, II, III, 1949–1958; An Ultraviolet Multiplet Table, *Nat. Bur. Stand. Circ. 488,* 1950–1962).

[23] R. L. Kelly, Vacuum Ultraviolet Emission Lines, (UCRL 5612, Univ. California Lawrence Rad. Lab. 1960).

[24] V. Kaufman, L. J. Radziemski, Jr., K. L. Andrew, *J. Opt. Soc. Am.* **56,** 911 (1966).

[25] V. Kaufman and J. F. Ward, *Appl. Optics* **6,** (Jan. 1967) and *J. Opt. Soc. Am.* **56,** 1591 (1966).

[26] K. B. S. Eriksson and H. B. S. Isberg, *Arkiv Fysik* **23,** 527 (1963).

[27] R. A. Sawyer, *Experimental Spectroscopy* (Dover, New York, 1963) 3rd ed.

[28] G. R. Harrison, R. C. Lord, and J. R. Loofbourow, *Practical Spectroscopy* (Prentice-Hall, Englewood Cliffs, N.J. 1948).

Author Index

Underscored numbers indicate the complete reference.

Subject Index

Abnormal glow, discharge, 133, 134
Absorptance of air, 1, 85
Absorption cross section, definition, 85
  O_2, 204, 205
  rare gases, 270–273
Absorption edges, 184
Absorption spectrum, argon, 121
  helium, 121
  krypton, 112
  neon, 121
Acetone ($[CH_3]_2 CO$), ionization potential, 255
Aluminum (Al), absorption coefficients, 189
  aging effect of reflectance, 38
  photoelectric yield, 230
  reflectance, 35, 36, 38
  transmittance, 188, 189, 191
  window, 252
Aluminum oxide (Al_2O_3), absorption coefficients, 201–202
  reflectance, 35
  thin films, 201–202
Amperite ballast tubes, 148–150
Angular dispersion, 10, 11
  compared to resolving power, 49
Anode dark space, 132
Anode glow, 132
Anodizing, 201
Anthracene, fluorescent emission, 224

Antimony (Sb), transmittance, 189
Apiezon grease, vapor pressure, 88, 89
Arc discharge, 133
Argon (Ar), absorption cross section, 270, 273
  absorption spectrum, 121
  continuum emission spectrum, 103–105
  ionization potential, 131, 254, 270
  line emission spectrum, 142–143, 153–155, 171–173
  resonance line, 131, 142–143
Astigmatism, 14
  curvature of image, 15, 16, 73
  image length, 14, 15
Aston dark space, 132
Atlas of emission spectra, 138, 175
Autoionization, 265
  structure, 270–273

Ballast resistor, 131, 132
Ballast tubes, amperite, 148–150
Barium fluoride (BaF_2), transmittance, 181, 182
  irradiation effect on, 184
  temperature dependence of, 182
  window, 252
Bendix multiplier, channel, 237
  magnetic, 236, 237, 293
Benzene (C_6H_6), ionization potential, 255